TAKE
BACK
YOUR
POWER

STORIES COLLECTED
BY MARLA MCKENNA

Take Back Your Power

Contributing Editor: Marla McKenna

Associate Editor: Griffin Mill

Contributing Authors: Laurie Raé Graham Bennett, Karen Carr,
Donna Drake, Tracy Feldman, Kathleen Gara, Manette Kohler, Sharon
Maniaci, Mary Markham, Peter Marks, Marla McKenna, Scott Melesky,
Adam A. Meyers, Natalie M. Miller, Ann Newman, Markos Papadatos,
El Pellegrino, Stefan Rybak, Connie F. Sexauer, Sandy Shelton-Wysocki,
Alisa Stevens, Abigail Tate

Cover Design: Marla McKenna

Interior Layout: Michael Nicloy

Take Back Your Power

ISBN: 978-1-957351-71-1

NICO 11 PUBLISHING & DESIGN
MUKWONAGO, WISCONSIN

Be well read.

www.nico11publishing.com

Printed in the United States of America

Connie Sexauer
September 14, 1945 – August 16, 2024

    This book is dedicated to the memory of our friend and coauthor, Connie Sexauer, whose strength, wisdom, and presence continue to guide us. Connie's life was a tapestry of love, education, and community service. She had always been a strong emotional support to her friends and family as well as helping numerous students realize their full potential. She was so excited about our new book. I wish she were here to see it. Though she is no longer with us, her spirit lives on in every word, every page, every book she has written and every meaningful conversation she has shared. Thank you, Connie, for the countless lessons you imparted, for showing us the true meaning of strength and compassion, and for leaving a legacy that will never fade. You are deeply missed.

TAKE
BACK
YOUR
POWER

Table of Contents

Foreword

By Tianna Madison

None of us is helpless. I'd like to begin there because we come into this world at birth—a bundle of potential and possibility—and over time, we adopt limitations from our environment, our relationships, or our beliefs passed down to us. We go from thinking we can do anything (remember that beautifully stubborn and confident toddler phase?) to wondering if we even deserve to be here … Hello Imposter Syndrome!

My memoir, *Survive and Advance*, which was beautifully edited by the author of this profound collection, proves that neither I nor you are unique in walking that path. But here we are, still standing, and still proving that none of us is helpless.

You always have power. One of the most insidious side effects of traumatic experiences and adverse events is that we rarely give ourselves credit for surviving. Instead, we view what we had to go through as something discouraging and disempowering.

Trust me on this; as a former professional track and field athlete and two-time Olympian with three Olympic gold medals I have dozens of these experiences I can use as evidence. In 2016, the year of the Rio Olympics, I was having the worst competitive season of my career. I lost every single competition. As each competition drew me closer to the Olympic Games, I began to wonder how I would reach my goal of becoming an Olympic Champion if I couldn't even win the smaller competitions—often featuring the same athletes who would compete at the Games. At the same time, I held fast to my personal mantra: "Everything is as it should be." My logic was this: if I believed that my steps were ordered, or predestined by fate, God, or some other divine energy, then whatever I was going through had a purpose. It then became my job to ascertain what that purpose was—or at the very least make it productive.

So, what was the "purpose" of these losses? With every loss I learned what not to do! I would enter the next competition knowing what mistake not to make. And although I made new ones, new errors, by the time I stepped foot on the runway at the track and field stadium in Rio De Janiero, there was nothing left for me to do but win. Armed with all the knowledge I amassed from all the losses I took, I stood empowered and confident and won the Olympic Games.

Now, I know what you're thinking; not all of us are Olympians and will experience taking our power back in such an obvious and dramatic fashion. I'm also a domestic violence survivor who after years of physical, verbal, financial, and sexual abuse, finally decided to leave that situation. It's been almost eight years, and I have not looked back.

I am also a mother, to a very independent and active toddler boy. When he and I travel, I am heavily dependent on our wagon stroller, its sturdy seat belts, and its storage capacity to get us from point A to point B; as it's largely just the two of us traveling together. On our last trip, I noticed—after folding the stroller and leaving it as directed at the foot of the jet bridge—that I only had one of the two wheels that pop off when I made it back to my seat. Somehow and some way between the boarding door of the plane and our seat just 15 rows back, one of the wagon wheels went missing! With a layover in Chicago, I began to quickly panic and spiral: *How am I going to carry him and our carry ons through the O'Hare airport?*

Do I leave the wagon here since it's absolutely useless now?

How did this happen?

Why am I so stupid?

Why did this have to happen NOW?!

I rang the flight attendant's call button and informed them of what happened. Politely they offered unhelpful reassurances, "It's here somewhere," or "I'm sure you'll find it."

I wanted an announcement made over the PA system. I wanted folks on hands and knees helping me search for a wagon wheel that just had to be on the plane. But none of that happened, and that help wasn't offered. So I sat in my middle seat, looking at the back of my son's curly-haired head as he looked out of the plane's window pointing out all the things he could see. I took a deep breath and repeated my mantra, "Everything is as it should be."

These words forced me to look at what I could do and turn away from what I felt I couldn't do. I realized that because it's a wagon stroller I could pull it. If I put my son on the side with the permanent tires, I could add the one wheel and pull the wagon to the next gate. *Okay*, I said to myself, *I have a plan now.* Relief began to ripple through my body.

Next, I received a text from United informing me that my arrival gate would be C17, and my next flight would be departing from C18. Visions of chaotically navigating Chicago O'Hare with three of four wheels dissipated as I realized I only had to walk one gate over! That also meant I had a little more time to look for the wheel once the plane landed. But I knew that even if I could not find it, all would be well. I did not find it; that missing wheel will forever be a mystery. I popped the one wheel on to the front of the wagon, and to my surprise, the wagon stood as it normally would, as if it weren't diminished structurally at all!

I laughed to myself as I realized that this is what so many of us do, jump right to the worst case scenario, look to others to save us (and are often disappointed by their inability to do so), before even giving ourselves the opportunity to do what we have been 100% successful at doing so far—figuring it out.

It does not really matter what the "it" is. We are all here today because of our ability to keep putting one foot in front of another. So this helplessness we often feel, is at best an illusion, at worst a downright lie. Remember, feelings are always real, but not always correct. In the face of any situation, you will always be able to ask yourself, "What can I do?" And there will always be an answer!

In the following pages, you'll hear from women and men, of all walks of life, telling their stories of taking back their power. I hope these stories are a mirrored reflection for you and your ability to do the same. None of us is helpless, all of us are powerful, and if you've given any of your power away, it's time. Time to take back your power!

Tianna Madison was born and raised in Elyria, Ohio, just west of Cleveland. She was a nine-time high school state champion, tied with Jesse Owens for most state titles. She earned a full academic and athletic scholarship to the University of Tennessee where she was a two-time South Eastern Conference Champion (SEC) and a two-time NCAA champion. Tianna is also a two-time Olympian and three-time Olympic gold medalist. She won her first gold medal by leading off the world record-setting women's 4x100m relay team at the 2012 Olympic Games in London, while also taking fourth in the women's 100-meters in London.

Four years later at the 2016 Olympic Games in Rio de Janeiro, Brazil, Madison added two more gold medals to her collection. She set a new personal best with a mark of 7.17m/23-6 to win the women's long jump competition and made it back-to-back Olympic titles by again leading off the women's 4x100m Olympic champion relay team.

Tianna is also a three-time World Champion in the long jump, taking home the title at the 2005 World Championships in Helsinki, Finland, and again at the 2015 World Championships in Beijing,

China. She also won the 2006 World Indoor Championship in the long jump in Moscow, Russia. She also has earned three bronze medals at world championships during her career.

In addition to her successes on the field, Tianna is a trauma-informed yoga and meditation teacher, advocate for various social justice initiatives, and has a social work degree from the University of Tennessee. She now uses the full range of her education as a track and field coach at San Jose State University.

She is passionate about using her lived experience and education to help others become high-performing individuals and teams while safeguarding their mental health.

Tianna speaks in front of both domestic and international audiences. Most notably a panel during Super Bowl weekend on behalf of women in sports, the College Football Hall of Fame in Atlanta, Georgia, in Oslo, Norway, for a SHE Conference on the topic of resilience before hundreds of attendees, and at the Cannes Lion Festival France on a panel for disparities in healthcare and the black maternity health crisis. Tianna is a passionate advocate for Black maternal health, driven by her personal experience. She nearly died during childbirth, and her son was born at just 26 weeks, requiring 73 days of care in the NICU.

Tianna is a credible thought leader having been called upon by the United States Olympic Committee to discuss racial bias on a year-long committee in addition to being published in *Yoga Journal*, *The Telegraph (UK)*, *The Los Angeles Times*, *The Washington Post*, *The Irish Times*, and *Newsweek* on a variety of subjects. She is also the author of her riveting memoir *Survive and Advance*, a tale of a life spent running.

Now that Tianna is retired from athletics, she channels the same energy, lived experiences, and social work background into making impactful change in the communities she is a part of. Tianna is dedicated to leveraging her platform and expertise to advocate for equitable healthcare, social justice, and mental health awareness. Her commitment to creating positive change and uplifting others remains unwavering as she continues to inspire and lead by example.

Introduction

It is with deep gratitude that I take a moment to thank God for uniting these motivated authors and allowing me the opportunity to collect and share their stories. I'm grateful for each one of them and the courage they've shown in expressing their vulnerabilities and for the strength they've found in taking back their power.

A heartfelt thank you to Tianna Madison for contributing her inspiring foreword. It's an amazing honor to have her voice included in this book.

To Mike Nicloy of Nico 11 Publishing for all of his hard work in making *Take Back Your Power* a reality.

***

As you read this book, you will find an abundance of useful tools and encouragement to help you reclaim your power. The common thread woven through every chapter is this: while an external force may have contributed to the loss of your power, it is an internal force that will guide you back to it.

That internal force is self-love. When you nurture your soul and your well-being, you free yourself from external influences that may diminish your confidence and happiness. Self-love empowers you to prioritize your needs and make decisions that reflect and align with your authentic self. When you fully embrace and accept yourself, you regain control over your life and choices. True inner peace begins the moment you choose not to let anyone or anything—be it a person, an event, or even illness—control your mind, heart, body, or soul. When you take back your power, you refuse to let external factors control your internal being.

If you search the internet for these key words: take back your power, you will find countless motivational quotes. I've shared a few of them here.

"One day she remembered that the only person who could make her happy was herself! So she took back her power, reclaimed her place in the world and shined like never before!"

- Unknown

Just as a flower blooms when watered, you too will blossom if you consistently nourish yourself with love, positive thoughts, supportive people, kind words, healing, patience, hope, open-mindedness, acceptance, self-belief, and self-care.

"If you are waiting around on somebody to save you, to fix you, to even help you, you are wasting your time. Because only you have the power to take responsibility to move your life forward. And the sooner you get that, the sooner your life gets into gear."

- Oprah Winfrey

Many of the authors have found therapy to be helpful which has allowed them to strengthen their sense of self.

"Therapy is about understanding the self that you are. But part of getting to know yourself is to unknow yourself—to let go of the limiting stories you've told yourself about who you are so that you aren't trapped by them, so you can live your life and not the story you've been telling yourself about life."

- Lori Gottlieb

They also recognize their strength comes from relying on a higher power—whether that be God, the universe, fate, or something else. Their faith is a strong guiding force to taking back their power.

"Be still and know that I am God."

- Psalm 46:10

I now invite you to read *Take Back Your Power* and embrace an open mind as you page through these inspiring stories shared by brave people who faced trauma, advocated for themselves, found happiness again, and reclaimed their power.

- Marla McKenna

SANDY SHELTON-WYSOCKI

"There is no heart for me like yours."

– Maya Angelou

How Destiny and Our Choices Can Change Our Lives

The Early Years

Growing up, I always felt the need to excel. I believed that being great at things would somehow win me more love and affection from those around me. *What a crock of shit!* I never knew why I felt this way, but it was ingrained in me from a very young age. Every decision I made was based on making others happy. As the years went by, it all became very clear why I did the things I did.

Traditions & Culture

I was born in Seoul, South Korea. My father, an American soldier, was stationed there when he met my mother. They fell in love, got married, and had two children. My mom is Korean and my dad is Caucasian. In Korea, kids like my sister and me are called "honhyeol"—mixed-race children. Although multicultural families have been part of Korean communities for decades, they are often looked down upon due to the ideology of racial purity.

This mindset dates back hundreds of years. In the early 1900s, when Japan was attempting to colonize Korea, they strongly believed in the concept of a unified bloodline and culture. Fast forward to the Korean War and the post-war period, the president of Korea continued to promote the ideology of racial purity. However, during this time, there was also a rise in mixed-race children born to American soldiers and Korean women working on the army base.

Post-Korean War, the government was trying to promote racial purity while the increase in "GI babies" created conflict. Because these babies were mixed race, many were sent overseas and given up for adoption. The president referred to the GI babies as "garbage" and a problem that needed to be taken care of. During this time, there was little regard for and much negativity toward mixed-race children. For some, this mindset still holds true today.

We eventually moved from Korea to Chicago, Illinois. Growing up, my sister and I were raised with a blend of Western and Korean traditions. Both of our parents loved us very much, and they lived their lives for us. My father, being from the South, instilled traditional Southern values and discipline. My mother adhered to strict traditional Asian culture, placing heavy expectations on us like achieving great grades, securing well-paying jobs, and buying homes in affluent neighborhoods. Over the years, it has been easy for me to see disappointment in her eyes, likely fueling my constant need to please others.

Korean culture emphasizes respect for parents and elders, the importance of appearances, social status, and academic achievement. Most parents aim for their children to become more educated and successful than they were.

These facts are important to know as I continue my story.

Defining Moments in Life

Do you ever think about defining moments in your life? In September 1982, I started my freshman year of high school. There was a particular boy who sat behind me in homeroom, and I was immediately drawn to him. Every day, I looked forward to our talks and sharing life's stories. We crossed paths frequently in school since he was a football player, and I was on the pom squad. It was a tradition on Pep Rally Fridays for the poms and cheerleaders to wear a football player's jersey. The week leading up to Pep Rally Friday, with my best friend Karen in tow, I knocked on the football locker room door where all the players were after practice. We knocked until one of the players opened the door. In that moment, I mustered up the courage to ask him for his jersey. That day changed our lives forever. He became *"the one."*

During the next three years of high school, he carried my books to class and held my hand. We talked, we laughed, and it wasn't long before we fell in love. No one ever made me feel as safe as he did, and no one ever loved me as much.

He loved old cars and drove a white 1970's Dodge Challenger. We often listened to Led Zeppelin's *In Through the Out Door* on the car's cassette player. In December 1985, he gave me a diamond promise ring for Christmas. We were young, in love, and believed our relationship would last forever. We dreamed about our future together.

I still remember the look on his face when I told him I had been accepted into Illinois State University. He was my biggest supporter in everything I did; he always believed in me. But I saw the sadness in his eyes knowing I would be leaving.

It was now our senior year, and shortly after my college acceptance, he broke the news that he was going to drop out of high school. "Wait, you're going to do what?" I asked, shocked. Years later, I learned about the many extenuating circumstances that led to his decision. At the time, I didn't know all the facts. There were many things happening in his life outside of our relationship that I had no knowledge of.

So, just like that, I broke things off. The man I loved, the man who made me happier than anyone, the man who made me feel safe—I just ended it. Having a relationship with someone who would go to college and follow the same path had been ingrained in my mind—someone who would have a great career. Partnering with a man who went to college and obtained a successful job that would lead to a successful career—this is what would make my mom proud.

Remember academic achievement being an Asian cultural thing? This would not be the last time it would come up.

We had a painful breakup. Later that year, his best friend began to date my sister. This made things especially difficult after the breakup because we had many run-ins with one another.

Reconnecting-Our Love Never Died

During my time away at college, it didn't matter how many years had gone by; each time we were in the same place, our eyes would lock, and we were right back where we started. I began working in central Illinois in 1990 and continued there until 1996. During that time, he and I reconnected once again.

In 1993, when my father passed away, he was there. He was always the one constant in my life. During the summer of 1995, while visiting my sister, who was now married to his best friend, I reconnected with him once more. This time was different. He was different. He had matured and was the best version of the man I knew growing up.

We spent the afternoon together. We walked, held hands, and talked about the beautiful life we could have together. No matter how many years passed, we never left each other's hearts. He loved me like no other and never gave up hope that one day we could be together.

Getting Mommy's Approval

After all these years, I decided I was ready to have a life with this perfect man whom I had loved my entire life. But first, I needed my mother's approval. I so desperately wanted her approval in everything I did. Perhaps it was the culture I grew up in. Perhaps it was the need for approval and the need to please her. Perhaps it was the need for acceptance.

So, I waited for the right moment. One day, as we were driving in the car, I decided to break the news to her. I told her we were planning a life together. As I began to tell her that the man I loved for so many years still loved me, she exploded. I've never seen my mother so furious and upset. "He isn't right for you. He doesn't have a college education. What kind of life will you have? No!" She screamed. She did not approve. Her response once again broke me. And just like that, her words made me turn my back on the man I had loved my entire life. I once again walked away and left him behind. But this time, I ran. I ran as far away from him as I could.

Moving On

It had been over a year since I had spoken to him. In the later months of 1996, I met someone else at work. We began to date, and shortly after that I left my job and took a position with another company. During that time, our relationship began to escalate as we kept in touch and spent a lot of time talking. Being an open book, I shared all my hopes and dreams, what I wanted out of life, and the little things that were important to me. Slowly, this new man in my life began to do all the things that mattered to me. He was seemingly romantic, we enjoyed the same interests, and we spent time together with our great friends. He also believed in the Lord. It was almost too good to be true. He came from a good family and lived in a wonderful suburban area. I thought to myself, *Would my mother finally approve?*

By 1998, I was approaching 30. All my close friends were getting married, and I wanted the same and to become a mom. And he checked all the boxes. Executive-level job? Check. Came from a good family? Check. Lived in a prominent neighborhood? Check. I think this time Mom would approve.

In August 1998, three days after my 30th birthday, I got married. We planned a large wedding with all the bells and whistles. My mother couldn't have been happier. It was big and fancy, something she could brag to her friends about.

I'll admit the first few months were not bad. We enjoyed being newlyweds, but even then, in the back of my mind, I always wondered if the boy who captured my heart when I was 15 was okay. *Where was he? How was he? What was he doing?* My sister later informed me that the news of my wedding devastated him. Up until the day I got married, he was still waiting for me—waiting for my call for us to finally be together.

Once he found out I was married, he finally gave up on me. He gave up on the love he had and decided he needed to move on with his life. In June 1999, he finally married someone else.

God Works in Mysterious Ways

By the time he and I married other people, we had very little communication. I did everything I could to avoid seeing him, even though he was my brother-in-law's best friend. I had young nephews, and over the next 20 years, we would occasionally run into each other at family birthday parties and summer BBQs. I did my best to avoid all contact and declined almost all family celebrations just to avoid seeing him.

We were both married to other people and had our own families. On the rare occasions we did end up in the same place, we would do a good job of being cordial with a quick, uncomfortable hello. Both our spouses were aware of our past. In fact, I once told my husband that I would always love him and only wanted the best for him. I knew the woman he married.

It's funny how God works in our lives. I believe we must go through hard times to find God's gifts in the pain we endure. He raised two stepchildren and had the son he always dreamed of having. They needed him as much as he needed them. That was the gift that God blessed him with. He was married for 17 years and divorced in 2016.

During my marriage, I also raised two stepchildren and had a beautiful son—the baby I always dreamed of having. My marriage ended in divorce, finalized in October 2018.

After nearly 20 years, I often ask myself *why did I stay so long?* I like to break the evolution of my first marriage into stages. The first seven years, I spent all my time working and building my career. During that time, I gave myself to my job and fell into a phase of robotically settling into life. I lived in a nice neighborhood, had great friends and neighbors, and a career I loved. It almost fulfilled me. But the void in my life was being a mom. My then husband had promised me before we had gotten married that we would have children right away. We even joked about having the first millennial baby. It's one of the reasons I got married so quickly. Shortly after our wedding, he decided he wasn't ready for any more kids. He also decided to leave his long successful career and start another. He sat me down and asked, "I am your husband, will you support me?" Little did I know what was about to happen.

Being the people pleaser that I was, I went along with it. I focused my time and energy on trying to be the best stepmom I could and building my career. His new career did not work out, so he decided to move on from that career path. He never worked again. As a divorced dad with only visitation rights with his children, he spent most of his days on the internet plotting revenge on his ex-wife. I witnessed this and did nothing. I felt powerless to what had become a new normal in my life. But the thought of having a baby was always at the top of my mind. I ignored the unhappiness that plagued me and focused primarily on my job, working countless hours surrounded by those I loved in the workplace. I advanced my career.

Seven years went by, and periodically I would bring up the baby conversation only to be shut down again and again. So, I decided to leave. I signed our house over to him, packed my things, signed a lease, and was ready to move on with my life. I was gone for three months.

So many people in my life have asked, "Why did you go back?" The answer was easy. He promised me a baby. So, I went back. Within six months, I was pregnant. In July 2006, my beautiful son was born. Once I had a baby, I was in baby bliss and solely focused on raising my child.

The next phase of my life was easy. I focused on my son and my career, but my heart was empty. We struggled financially as I was the only income supporting our family while living in a very affluent neighborhood. My husband controlled everything—every aspect of my life, including my finances. It got to the point that every two weeks when my paycheck arrived, I turned it over to him and had no money. He said he would be responsible for paying bills, and that's the last I saw of the money. But I had my son.

By 2012, I once again found strength, or so I thought, to leave. I sought out an attorney to weigh my options. What I was told was devastating. According to the attorney, it was likely that my then husband would retain sole custody as he was the primary caretaker. Because he was the stay-at-home parent and I was the working mom, the law was not in my favor. Hearing this news was devastating. So, I decided at that moment that I would stay and try all I could to make things work. I could not leave my child. He was only six years old.

I read self-help books, sought marriage counseling, and even immersed myself in church, looking for the happiness I longed to have in my life. It was a temporary band-aid. I was able to get by for the next five years. But by this point, he controlled everything in my life—my friends, my finances, who I spent time with, what I watched on TV, the music I listened to—everything.

Over the years, my close friends saw it. They suspected it, but I played it off like I had the perfect life and perfect marriage. In my head, that is what I longed for. I was too embarrassed and too ashamed to tell anyone of the misery and torment I was going through. So, I remained silent and put a smile on my face. Remember Asian culture—appearances are everything. I was powerless to a man who took everything from me: my confidence, my self-esteem, and my finances. I lost all power to this man. And I allowed it to happen.

Enough Is Enough

In the summer of 2017, I decided enough was enough and filed for divorce. Before this, I had been very much a people person, surrounded by friends and socially active. But during the process, I fell into a dark hole. I closed myself off from everyone around me and completely shut down. The divorce drained me financially, leaving me nearly penniless. We took a huge financial loss on the sale of our home, and I had to pack up my belongings and put them into storage. Once the house was sold, I had no place to go except to stay with my family in another state. I thank God for my family who took me in. It's hard to be a successful career woman and mom while at the same time pretty much homeless. We also decided it would be in the best interest of our son not to uproot him from his school, so he stayed with his dad during this transition, remaining in the school district. I cannot even begin to describe the gut-wrenching pain this feels like. For any parent going through this difficult transition, not having your child with you is a pain like no other.

Do you ever wish you could turn back time? Stand up for yourself? Do things differently? It took nearly five decades for me to realize that I gave away my power at a very young age by putting others' needs before mine.

I have come to understand that not only did my cultural upbringing and the way I was raised affect me emotionally and psychologically but also how being "honhyeol"—mixed race influenced how certain people in my life viewed me. It instilled a mindset of constant people-pleasing and the need for acceptance.

I often ask myself, *What would my life be like had I done things differently? What if I didn't feel the great need to be accepted? What if I put my happiness first? Would I have done things differently in my life? What if I had put my heart first and just said yes when I was 27 years old? What if?*

Fate, Power, and Eternal Love

During that year, as my family gathered to plan my nephew's wedding, the concept of "Inyeon" started to surface again. Inyeon is a Korean word that means "bond" or "relationship," representing the strength of the connection between two people and how their paths are destined to cross.

I believe that "Inyeon" once again brought me back to the first man I ever loved. He had always been like an uncle to my nephews and a significant part of their lives growing up. During the wedding planning, we reconnected. Nearly 40 years later, the love we shared as teenagers, then as young adults, and now as grown adults, had never died. I would like to say we found our way back to each other, but I think God and "Inyeon" placed us in each other's paths again.

The day he proposed, he said, "I have loved you my whole life, and there is no one I want to spend the rest of my life loving other than you. Will you marry me?" The only thought that ran through my mind was the happiness this man brought to my son and me. The love he has shown to both of us. The man who has loved me nearly his entire life. The man who makes me feel incredibly safe, appreciated, and loved. Without hesitation, I said "YES." We were married in June 2020.

If you had asked me several years ago how I thought my life would be today, I would have never imagined telling you that I have found joy, happiness, and a love beyond anything I have ever known. Life is funny. So many times, we pray for the things we think we want or need. The saying that God works in mysterious ways is so true.

Going through a divorce and losing nearly everything was the worst pain I could experience. But finding the strength to take back my power, to endure the darkness, and push toward the unknown led me to a place of true joy and happiness. True destiny. My own "Inyeon."

Sometimes you must ask yourself, *Are you in the right place for what you want, even when it seems impossible?* Don't be afraid to take back your power and find out.

About Sandy

Sandy was born in Seoul, South Korea, to a Korean mother and an American father. At the age of two, her family relocated to the United States, where she grew up in the vibrant community of Southeast Chicago, known as "The East Side." This unique upbringing allowed Sandy to seamlessly blend and adapt to both Eastern and Western cultures.

During her public high school years, Sandy met her true love. Although they went their separate ways, after several years, fate brought them back together many years later. Sandy went on to study at Illinois State University, where she earned a Bachelor of Science in Applied Science & Technology. Her dedication and achievements in her field have earned her a distinguished place in the College of Applied Science and Technology Hall of Fame.

With a career spanning 35 years in the retail industry, Sandy has demonstrated exceptional commitment and expertise. She currently resides in Northwest Indiana with her family, cherishing the moments spent together. Outside of her professional life, Sandy enjoys spending time with her family outdoors and indulging in her passion for gardening.

Her life story is a testament to resilience, love, and the beauty of diverse cultural experiences.

Sandy, her dad, and older sister, Linda (Korea).

Sandy, her mom, and sister, Linda (Korea).

Sandy's first birthday, wearing traditional Korean hanbok.

Sandy and Scott, 1985.

Sandy and Scott, 1986.

Sandy and Scott, 2018.

Sandy and Scott, 2020.

KATHLEEN GARA-MELVIN

"I'm an 18-years thriver, I call myself, not a survivor, because I feel like I'm in thriving my life even better than I ever have, and I want to encourage other women that are going through this journey that here I am."

– Olivia Newton-John

I Had Cancer, But it Didn't Have Me

My battle with cancer began in fall of 2004. It was time for my yearly mammogram, but this time it would be different. They found something that didn't look right and wanted me to follow up for more tests. They also wanted to do a biopsy on the lump that they had found. My doctor's office would call me with the results on Friday afternoon. I had to work that day, but I told them to call around 3 p.m. that afternoon when I would be on my break.

My husband, Dave, had gone up to our cabin late Thursday night after work. I told him to call me Friday night, and I would let him know the results. He called me Friday morning around 10 a.m., and we talked for a few minutes before I had to leave for work, and of course at that time, I didn't know anything yet. I went to work and waited to hear from my doctor's office. At 3 p.m., I received the call with the words that no one wants to hear, "You have breast cancer."

I was scared and started to cry, but I knew I had the strength to get through it. I had to wait until I returned home before I could tell my husband. He was supposed to call me at 9 p.m. The time passed, but he didn't call. When it got to 10 p.m., I tried to call him, but there was no answer. I wasn't too worried yet, as the cell service was weak at the cabin. I had to work early the next day, so I thought I would tell him later when I got home. But when Saturday evening came, I still couldn't get through to him on his cell. As Sunday came and went, I experienced the same unfortunate experience. I finally decided to call a neighbor who lived near the cabin. They said they would go check on him for me, as they could see his truck parked by the cabin, and the lights were on. A little while later, they called me back, and said they had knocked on the door, but there was no answer, so they looked through the window and could see him lying on the floor. They called 911 for help. I told them I would drive up right away. It's a two-hour drive, and as I arrived the sheriff and an ambulance were there waiting for me. My husband had passed away, and they thought it had probably happened on Friday. He had passed away without knowing I had cancer. My life unexpectedly changed overnight, and I never felt so alone.

The next few weeks were busy, not only with doctor appointments and setting up my surgery but also planning a funeral. This was all happening during the holidays. It would be the first year without my mom, who had passed away earlier in 2004, and now without my husband. I knew I had to be strong, and just get it all done; I did. The funeral was the beginning of December, and my lumpectomy was about a week later. Everything went well with the surgery, and they made me spend the night at the hospital because they knew I would be at home alone. When the nurse came in my room the next morning, I sat up in bed, and she asked, "Doesn't it hurt?" I was a little surprised that it didn't. I thought to myself, *this wasn't so bad.* After breakfast, I was able to return home.

During my follow-up appointment, the doctor put me on a hormone suppressing medication for five years because my type of cancer fed on estrogen. I had to take off from work for a few weeks after the surgery, but soon I was healed. I was ready to return to work and back to my life. Little did I know, this would not be the end of my cancer battle.

Spring came, and I continued to heal, from both my surgery and the loss of my husband. The summer before, I had asked my husband to remove a rose bush that was no longer producing roses. Instead of throwing it away, he just moved it to another spot in the backyard. The spring after he died, the rose bush began to grow again, but this time it was filled with beautiful blossoming roses! I knew it was a sign that he was still with me.

During the summer of 2005, I began to date; however, none of the relationships were serious. It was just nice having a companion to enjoy things with. Then on the seventh anniversary of my husband's death, I received an email from a guy I had known in high school. *Was that a sign from my husband just like the blooming rose bush had been?* Chuck had lost his wife to a different type of cancer. We emailed each other for a few days, and then I gave him my phone number so we could talk. I had sent him a message a year before, on a Classmates website, but he hadn't seen it until now. We decided to get together for dinner and talk about what we had done since high school. I had only seen him once since high school, at our 25-year class reunion. We had a nice time together at dinner and decided to go out again. Soon we were seeing each other often, and it eventually became more serious. I hadn't known that he had a crush on me back in high school. Meanwhile my five years on the hormone suppressing medication had ended the year before, and in 2012 my cancer would return.

This time the cancer was growing in some lymph nodes in my neck. I went for a biopsy to confirm it and then had another surgery to remove the lump. I also had to go back on the medication because the same type of cancer had returned, only in a different place. Chuck was with me through it all, and our relationship continued to get more serious.

The next year for my birthday, he bought me a ring, and we became engaged. We got married three years later, in 2015. It was about 50 years after we first had met, and things were going well. We had started going to concerts together, and one of the artists we enjoyed seeing was Rick Springfield. We both liked his music and traveled to other states to see him perform. There were several fan trips Rick did, and we went to a few of them. Two were in Florida, and one was in the Bahamas. We also went on a cruise that Rick hosted. It was the first cruise either of us had ever been on. We met

Rick and began making friends with many of his other fans. We even ran into Rick and his band several times after other concerts, either at a bar or restaurant at a hotel where we all happened to be staying. It was a lot of fun getting to know other fans and the guys in the band. But the fun was about to change again.

In 2017, it was time for another mammogram, and once again there was something that didn't look right. There was more testing, and another biopsy was done. My breast cancer was back again. I later found out it was a different type than the first two times. It was not hormone positive, but it was something called HER2 positive. The medication I was taking didn't work on that type of cancer, so I had to go through more surgery.

This time the cancer had spread too much; I would need a mastectomy. My doctor gave me the choice of a single or a double mastectomy. Since I had already had cancer in my left breast, and now my right, I decided to do the double, but this time I wasn't alone. My husband, Chuck, was with me through it all. The surgery took about six hours. The breast surgeon removed one breast, and then the plastic surgeon started working on putting in the expander, while the breast surgeon worked on the other one. My husband waited in my hospital room and was there for me when I woke up. I was only supposed to stay overnight, but I became very nauseous from the anesthesia. I would sleep for a little while and wake up feeling nauseous again. The next day I still wasn't feeling very well, and all I ate was broth and gelatin. I had to stay another night at the hospital because they wanted me to be able to eat regular food before they would send me home.

On Friday, I was finally able to eat breakfast, and then I went home. For the first few days, the pain from the surgery was intense at times. I couldn't do too much around the house, but my husband helped by doing the cooking and laundry, etc. We had bought tickets to a concert before I found out I had cancer again, and only five days after my surgery, I told my husband I wanted to go to the concert. I said there was no reason not to because the pain wasn't as bad anymore, so he put me in a wheelchair, and we went. Even though I was sore and still healing, I wasn't going to let that stop me! We enjoyed the concert, and I just sat in the wheelchair. We also had planned and paid for a trip to Vegas, months before my surgery, and I told my doctor I was still planning to go to that too which was only three weeks later. He said it should be fine if I didn't do any heavy lifting and was careful because nothing was wrong with my legs. We went and had a great time and saw several shows.

In the weeks after my surgery, my friends sent me cards and get well wishes. My friend Jamie came to visit and brought me some cards that another Rick friend, Linda, had gotten signed by Rick Springfield and his band members. I also got some drumsticks from the drummer. I had several doctor appointments in the weeks that followed. The expanders that were put in during the surgery, to stretch my skin to get ready for the implants, had to be filled with saline almost weekly for the first two months My oncologist wanted me to start chemo and some other traditional treatments, but I decided against it. I found an herbal protocol that I wanted to try first. I didn't like all the side effects that I could have from chemo. So, I did it my way, and now six years later, my HER2+ cancer has not returned, even though that type of cancer can be aggressive and return, usually in the first two to three years. Finally in August of 2018, my exchange surgery from expanders to implants was scheduled. That surgery was much easier than the mastectomy. I healed well, and my life went back to normal once again.

Around February 2021, I noticed that I would get short of breath easily. Just walking to the mailbox would make me tired. I went to see my regular doctor, and she then sent me to see a pulmonary specialist. That doctor did several tests and X-rays. The X-rays showed a spot on my left lung that needed a closer look. I had to get a CT scan which also showed fluid on my lung. They also did a biopsy, and it showed my hormone positive cancer had started growing once again. I had stopped taking my hormone medication for a few months, and that let the cancer start growing. The doctor then sent me to see my oncologist. I was lucky I didn't need surgery, but I did get the fluid drained, and the oncologist changed my medication to Tamoxifen. After a few months, I had another CT scan done, and it showed the cancer had stopped growing. For now, I am doing well once again. I will continue to take my medication, and not let the cancer, or fear of it returning keep me from living my life. I will continue to go on trips and see concerts, with my husband and friends Jamie, Marla, Penny, and Jenn. I take back my power by moving forward with my life and not letting cancer defeat me. I'm not only a survivor, but a THRIVER! In August of 2023, I had another CT scan, and everything was great. No new signs of cancer, and the spots that were on my lung have decreased in size.

Unfortunately, in 2021, my husband was also diagnosed with cancer, and he had to have surgery and chemo. He was doing alright during the holidays but had to have another CT scan in February 2022. It was not good news. His cancer had spread, and he would have to do chemo again. He continues with his treatments and started a new medication at the end of 2023 which has slowed down the progression of his cancer now.

But that's another story and not mine to tell.

I dedicate my chapter to my loving husband, Chuck Melvin who passed away February 22, 2025 after his 4-year battle with cancer.

About Kathleen

Kathleen was born in Milwaukee, Wisconsin. She later moved to New Berlin with her family when she was 10 years old. It was there that she would meet her husband Chuck when she was just a freshman in high school. She didn't know at the time that they would one day meet again and get married 50 years later. In between high school and reconnecting with her former classmate, she married someone else and had two sons, Wayne and David. That marriage would end in divorce a few years later, and she would raise her boys as a single mom, without any help from their dad. She would eventually remarry in 1994, but that marriage ended 10 years later, when he passed away from a heart attack, the same day she would get her biopsy results showing she had cancer for the first time. She also lost her oldest son Wayne in 2002 due to an accident.

In 2012, she reconnected with her former classmate, and they were married in 2015. She and her husband have four grandchildren: Holly, Anakin, Charlie, and Emerson and two great-grandsons, Mason and Jackson. They are enjoying retirement traveling both in and out of the country, going to concerts, and spending time with family and friends.

Kathleen with several
of her grandchildren.

Chuck and Kathleen.

Kathleen and Chuck with
Rick Springfield.

Kathleen with
Rick Springfield.

ABIGAIL TATE

*"I am not what happened to me.
I am what I choose to become."*

– Carl Jung

Climbing Back Up the Hill

Growing up in my family, indestructible was one of the many words used to describe me. I felt I had all the power in the world. Every crazy stunt or dare I did, I always seemed to walk away without a scratch. Fear wasn't even in my vocabulary. I never shied away from holding snakes, riding the tallest roller coasters, or making the first move on a date. I was born with this relentless quality, and it was unwavering until my sophomore year of high school.

I started my day off like any usual morning. "Usual" as in, firing up my laptop for e-learning. My day dragged on as I eagerly waited to go sledding with my friends after class. It had been months since I'd seen them because of the COVID-19 pandemic, and I excitedly awaited our reunion. Finally, 3:15 p.m. hit, and I rushed into my snow gear and out the door. Soon, we all arrived at a nearby park. Seeing my friends for the first time in months filled me with happiness and content.

The three of us rushed to the top of the hill, stumbling on our way up, and then raced back down on our sleds. The thrill of the night went on as we traveled up and down the slope. As more time passed, the park became overly populated. Therefore, we were forced to sled on the opposite side of the hill. There was only one problem … trees scattered the land.

My friends and I were skeptical about our situation until we saw a young family cruising through the forest. I believed if others could do it, so could I. With this in mind, we made our way to the very top of the hill. I volunteered to go first. I sat on my sled staring down the long and steep slope that lied before me. With no hesitation, I fearlessly pushed myself off the top. As I traveled further and faster down the hill, I began to lose control. In an instant, my sled turned backwards, and I had no idea where I was going. My mind was racing, but I ultimately knew this wasn't going to end well. A few seconds later, I collided with the last tree that towered over the hill.

My vision went black, and my ears began to ring. I tried to call out for help, but nothing came out. I couldn't breathe. When my vision returned, my eyes set on the young family a few feet away. I started to panic. I needed to get their attention. A few moments later I was finally in their gaze. They rushed to my side and asked if they needed to call my parents. With the last breath in me, I struggled to get out the numbers, "911, 911, 911." I then collapsed onto my side. The family called the paramedics while my friends came running down to me. I laid there trying to stay awake. It was getting harder and harder to breathe. I started to consider the thought that this is what dying felt like. The minutes passed by while my pain increased. Somehow through all of my agony, I felt peaceful in a way, almost like accepting my fate. I was thankful I could feel how cold the snow was. I tried to focus on that to distract myself from what was really happening.

The sound of the sirens in the distance started to get louder as the ambulance got closer. It was odd hearing those sirens, knowing they were for me. It was almost as if I was hearing them for the first time. When the paramedics arrived, they put me on a board and carried me to the ambulance. The ride to the hospital seemed to be over in the blink of an eye. But with every stop, turn, and go I winced in pain. The pain didn't subside anytime soon either. I waited for hours in the emergency room until I could be taken for X-rays which concluded that I had bruised my ribs. I was confused. *How could all of my pain just come from bruised ribs?* But I had to trust the professionals and believe what they told me to be true. I was then sent home at 3 a.m.

A few days later, I had a check-up with my pediatrician, so they could assess my injuries and document their records. Everything was going according to plan, until one test confirmed that I had internal bleeding. The doctor immediately sent my mom and I to the nearest hospital in their system. However, this medical center was different from the location I was originally sent to. These doctors, instead, performed a CT scan to look further into my injuries. Compared to my original diagnosis, this scan showed that I had fractured four ribs and had five compression fractures in my spine. The doctor then explained to us that one of the spinal fractures was resting on my spinal cord. I couldn't believe it. The whole time I was at risk of being paralyzed from my shoulders down. I was terrified to move, in fear that I would cause my bones to shift even more. With this new knowledge, I had to be transported to Lurie's Children Hospital in Chicago—the only hospital in the area that could treat my injuries.

After staring at the ceiling of an ambulance for an hour's drive, we finally made it to the city. At Lurie's, I was first treated like an incoming trauma patient with doctors rushing in to analyze my condition. I stayed in the emergency room for a few hours with healthcare professionals running tests on me. Once 10 p.m. hit, I was admitted to the hospital and given a room on the 19th floor.

The next few days were a blur ... consisting of an MRI scan, meeting new doctors, and watching movies on the TV. I was still in a great deal of pain and hoped things would change soon. And they eventually did when an orthopedic surgeon gave me the option to undergo spinal fusion surgery. The thought of surgery didn't scare me at the time because I would have done anything to take the pain away.

Once my parents and I made the final decision to go through with the surgery, we met with the orthopedic surgeon, neurosurgeon, anesthesiologist, and surgical team, who would be in the operating room. My surgery was scheduled for the next day, and I could not have been more eager. At noon the following day, I was brought into the prep room. The surgery lasted four hours, but it felt like no time had passed at all. During this time, the surgeons implanted titanium rods and screws along my T2-T6 vertebrae to keep the bones secured in place. I'm not sure what I was expecting, but I believed that surgery would make me feel better instantly. With this procedure however, I would have to wear a brace around my thoracic region and neck for the next four months, along bed rest. I was discharged from the hospital the following day and was told to live by three rules for the next few months: don't bend, don't twist, don't lift.

I was relieved to be home at last. But all I could think about was, *Now what? After being in the hospital for weeks, how was I going to be able to do this on my own?* Luckily, I didn't have to. My mom graciously set aside all of her other responsibilities, in order to take care of me; she was my biggest supporter throughout this time.

After a couple of days at home, I started to develop a routine. My day began by waking up in my parents' room; I slept with my mom because I wasn't able to sleep alone. My parents periodically had to switch me from my back to my side throughout the night and give me pain medication. In the morning, they would have to lift me up from their bed and guide me to the bathroom where my mom would help me wash my face and brush my teeth. I couldn't do any normal tasks by myself. I felt useless—like a burden.

Throughout the day, I would do online school. It kept me busy, but it didn't help that I was almost a month behind. Everyday after my classes, I had tutoring to catch up. My self-esteem started to plummet because for the first time in my life, I needed help with school. My grades were something that I had always been proud of, and now I was struggling to raise them. I didn't share my situation with my classmates because of embarrassment, and I was anxious about what other people would think of me. I assumed other people thought I was slacking off or unintelligent because I withheld the full story.

I tried to find comfort in friendships, but that was yet another let down. Although they were supportive in ways, I was upset in their company. I was reminded of everything I was missing out on. We did everything together, but I began to fade in the background of their lives. Everything they shared with me was so exciting and interesting. I told them details about my situation that nobody else knew because I wanted to feel included. In return, I was laughed at. From then on, I stopped seeing my friends and didn't share my trauma with anyone.

I no longer felt indestructible. Being fearless was something I could no longer say about myself. I felt lost. I stopped enjoying the things that I used to love. I stopped playing my violin. I stopped playing basketball. I stopped listening to happy music. Everything that made me the person I was had disappeared. I felt trapped in my house and in my broken body. After a while, the days started to blend together. Repeating the same thing over and over again was not entertaining. There was nothing to look forward to as I woke up every day. My mind would wander as I sat on the couch. Most of the time, I would retrace everything that had happened in my accident. Sometimes, I wished it never happened. Sometimes, I wished my outcome was worse. My mind became a battleground between trying to heal myself and idealizing my pain. Negative thoughts flooded my mind, and I became severely depressed and suicidal. I prayed for these feelings to go away, but it seemed as if I wasn't in control of my own thoughts anymore.

Months went by, and as my physical self improved, my mental health declined. During the summer, I began to see my friends and go out again. I was happy with them, but that happiness never remained. I was constantly surrounded by people now, so I didn't understand why I still felt so alone. I felt different from other people. This increased my social anxiety, immensely. As a result, I started to have panic attacks when I would go out in public. I had lost a great deal of my confidence, and I convinced myself that people were judging me based on the way I looked. Once school started again in the fall, my anxiety only got worse. I felt nauseous before leaving for the day and was uncomfortable in my classes. I wasn't sure how to cope with all of these feelings, and I started to take my pain and anger out on myself.

This mentality of mine continued for months. I started to feel like I was living for other people and not myself. I grew a fear of being alone. I knew what I was capable of and didn't want an opportunity to make any decision that I'd regret. I struggled with wanting to be here, but the thought of hurting myself would only hurt the people I loved which made me reconsider my thoughts.

It wasn't until I had gotten to my lowest point and suffered long enough that I realized nobody was going to get me out of this but myself. From that moment on, I started to prioritize my health and happiness first. For me, this consisted of connecting with nature, cutting ties with toxic relationships, and most importantly, going to therapy.

I never imagined myself as the type of person to be in therapy. I hated talking about my emotions with people. I would put a smile on my face for others but struggle alone. Based on past reactions of telling others about my accident, I felt very guarded. However, I learned that I would never be able to heal unless I faced my emotions head on. While in therapy, I talked through my whole experience, how I felt at the time, and how I am feeling currently. I've learned a lot about myself through this process. I'm now able to identify my emotions, cope with my PTSD and anxiety, and overall become a better version of myself.

When I first started my recovery, my goal was to return back to who I was before; however, my mindset at that time wasn't going to allow me to make any progress. I was chasing something that was impossible to get back. It took me a long time to realize that I am a different person now. I struggled with this thought until I discovered that certain changes don't have to be bad. Just because I was originally in a negative situation didn't mean that any change tied to it had to be in the wrong direction. I was surprised that a lot of good eventually came out of my experience.

For instance, I learned a lot about my own identity. I had to spend a ton of time self-reflecting and rediscovering who I was after all the trauma I had gone through. I learned I don't always have to be fearless, and I can rely on others when I need to. Bottling up my emotions just caused them to multiply and worsen. Once I learned to open up, my negative thoughts began to minimize. If I didn't express myself, I would never have been able to get through this on my own.

During my depression, I lost my love for what brought me

happiness. Although, I've gained plenty of new hobbies that I would've never opened my mind to if it weren't for the circumstances I faced. Because I had tutoring for math three days a week, I began to love the subject. I loved it so much that I joined the math honor society at my high school, Mu Alpha Theta. And while I was sitting at home for months, I decided to explore different music genres that I wouldn't normally listen to. In return, I have grown to love alternative and rock music. It's true what people say, "When one door closes, another opens."

Furthermore, this experience opened my eyes to a world I had never seen before. Spending a month in the hospital allowed me to meet a variety of people with different professions. This instantly sparked my interest to work in the medical field, and radiologic technologists stuck out the most to me. I became fascinated with the technology and knew what it had done for me, so I wanted to help others in a similar situation to mine. I began looking for colleges in the area with this major and came across a few. Ultimately, I chose to enroll at Carroll University in Waukesha, Wisconsin.

However, my life isn't all sunshine and rainbows now. I still have my moments where I struggle with what happened in the past. It has taken me years to feel some sort of progress. I've come to learn that no matter how small the progress I make, it's still growth and is just as valid. If this experience has taught me one thing, it's that there is no time limit to the healing process. And slowly but surely, I am becoming that fearless, powerful, and indestructible girl again.

How I took back my power?

- Spending time with loved ones
- Opening up in therapy
- Giving myself mental breaks when needed
- Embracing change
- Finding the good in bad situations
- Learning to love all versions of myself

I'd like to dedicate my chapter to my loving mom, who has been with me every step of the way. I couldn't have taken back my power without you.

About Abigail

Abigail is an accomplished graduate from Antioch Community High School. She completed her secondary education with high honors and Cum Laude. During the past four years, she was devoted to helping her community and school by volunteering through National Honor Society and Mu Alpha Theta. In addition, Abigail has developed a love for music by participating in the orchestra program for 10 years. Because of this, she enjoys going to concerts with friends and listening to her favorite artists in her free time. Furthermore, she spends her time going on nature walks, coffee dates, and watching her favorite TV shows. But most importantly, she loves to spend time with her parents, sister, and her beautiful dog, Lola.

Abigail is continuing her educational career at Carroll University. She is pursuing a degree in the health sciences majoring in radiologic technology. Abigail is a member of the Medical Diagnostics and Imaging Club at school.

Abigail's high school graduation. Go Sequoits!

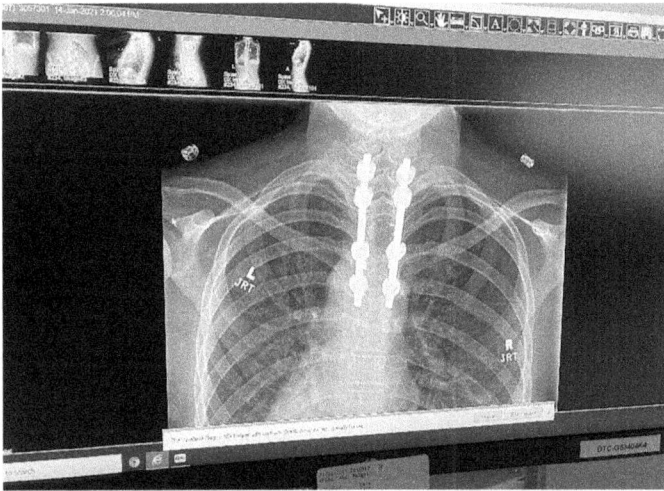

Abigail's chest X-ray with spinal implants from T2-T6.

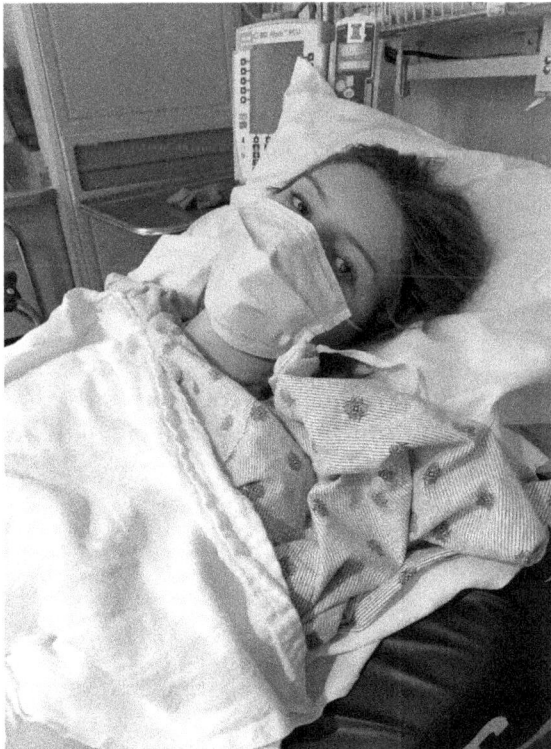

The first night of many hospital stays, January 5, 2021.

Forever thankful for her family. From left to right,
Abigail's mom, Penny, sister, Adalia, and dad, Daniel.

DONNA DRAKE

"Be who you are and say what you feel because those who mind don't matter, and those who matter don't mind."

– Dr. Seuss

Figuring Out Consistent Unique Strengths

Thursday, April 9, 2009, was the day my dream came true. That was the day my television show, *Live it Up! With Donna Drake*, now known as *The Donna Drake Show*, premiered. I was living the life that I had dreamt of in color; I was content with what I had accomplished. I've had the opportunity to meet and interview celebrities and travel to foreign countries. I've received accolades in the form of two National Telly Awards, a Greek Artemis Award, recognized by the Power Women of Long Island, acknowledged as an Amazon number one bestselling author for *Manifesting Your Dreams*, which appeared in *O, The Oprah Magazine*, and, most importantly, I have sponsored many charitable events while also using my platform to lift everyday heroes.

Then it hit me one day; *Why did I stop dreaming?* I'm so grateful for everything I have, as well as the array of circles I've filled with wonderful people who have appeared on my show. But if I didn't take action on that beautiful and vivid dream I had in color on December

28, 2008, I would still be the timid girl working promotions at WLNY-TV, hoping I would get the chance to peek into the newsroom. I took back my power by creating my own television show; however, that was just the beginning. I fueled my empowerment by utilizing my mantra F.O.C.U.S. (Figuring Out Consistent Unique Strengths) which I featured in *Manifesting Your Dreams.* I used this mindset to create my show, and now I invite you to journey with me onto my next chapter of empowerment.

As I strengthened my confidence to take over the old CBS/WLNY studio, where I once worked, I embarked on creating a NYS-qualified production facility in just two years, which is now officially known as Drake Media Studios.

In the past, I rented other studios to produce more content for the first 12 years of *The Donna Drake Show.* While in Salem, Massachusetts, on November 1, 2020, I received a phone call from the rental space company that was home to my show. They advised me that my hair and makeup room, as well as my office space would be unavailable to me moving forward. I had to leave all my lights and my set there. These roadblocks actually cleared the way for me to grow and advance to bigger opportunities in my career. I no longer wanted to be nomadic; I wanted to build a home with my crew—the family I'd been working with for numerous years. The search to find a new creative space began. After calling many studios and with no luck, I had an idea. *Why don't I try my previous work home, the old WLNY TV newsroom in Melville, since my show airs on a CBS affiliate, and they owned that studio?* Talk about divine timing; at that moment, Viacom had just purchased CBS. This beautiful grid-lit studio was available. I couldn't run to my car fast enough; I drove over and toured the space—talk about the serendipity of life! The studio was a shell of what I remembered it to be 20 years ago.

White spackling patches scattered the once beautiful wall, while loose hanging wires escaped in every direction through the missing ceiling tiles. This space was only a remnant of what used to be the booming WLNY TV newsroom set, but I fell in love with it. The gears started turning, and I began envisioning my new entrance hallway—a retro theme to match my childhood with a black and white Zenith television acting as a table on my set. I imagined my interns creating vibrant content in the now vacant spaces which were once edit rooms. I saw the potential in this available new, bigger space. Even though my imagination was limitless, my budget, unfortunately, was not. I needed to negotiate a deal to shave off half the footprint to afford the space. My intuition told me that I was brought back to this studio for a reason, and it would be mine, but I needed a physical sign. I looked at the dark "On Air" sign which was hanging above the entrance to the studio, and I said to myself, *I'm looking for a sign, and if I flip the light switch and it turns on, then that means that God wants me to take this space.* I hesitated, took a deep breath, and quickly flicked the light switch. And just like that, after all these years, the light illuminated.

Before I delve into how I used F.O.C.U.S. to take back my power and buy my studio, I'm going to explain the mantra's method. In 2017, my friend Lucy Rosen coordinated speakers nationwide for a TEDx Talk at Hilton Head in South Carolina. She asked me to apply. I sent them a mock TEDx Talk on Figuring Out Consistent Unique Strengths and was selected to present. When you are stressed, overwhelmed, or feeling insecure in an environment you're not comfortable in, F.O.C.U.S. helps you understand who you are and what makes you unique. Honing into YOU will fuel confidence, growth, and success in your life. I share this approach with people who aren't familiar with being on camera to gain the confidence they need to professionally represent themselves and their company.

F.O.C.U.S.

F.O. is "Figuring Out" who you are when you're out in life. What's your involvement with others or your community, and what are your roles in these activities? Snap a photo of yourself when you were successful at something.

C. is for "Consistent" things about yourself that people can always depend on.

U. is for "Unique," how you make a difference in the world.

S. stands for your "Strengths," a resource that assists you in your journey, serving as a stepping stone to where you are today and how to share them with others when they are no longer needed.

F.O.C.U.S. was born when I no longer had my mother, who was my biggest cheerleader, was divorced from my first husband, widowed at a very young age, and divorced again. Being a single mother and feeling very alone, I had no one to fall back on or provide support when I needed it, but I wasn't going to give up on my dream of creating my own television show. That's when I knew I needed to be resourceful and Figure Out my Consistent Unique Strengths. So, I decided to sell everything I owned and invest in myself. More than a decade later, I was on an even bigger venture. Despite how I had gone from almost nothing and ripping tickets at my local movie theater to my dream job interviewing celebrities from all over the globe, running a studio was an unprecedented feat. F.O.C.U.S. is not stagnant, it evolves when new questions and challenges arise. You can repeatedly use those same basic principles to produce new results for yourself.

I applied Figuring Out Consistent Unique Strengths to rebalance myself: I took a picture of myself back in this footprint, remembering who I was when I worked in the building. I visualized myself being successful when I was the director of creative services here.

The next step was "O," I thought about who I was and asked myself, *"I already have owned a company; could I be that entrepreneur? Could I run an actual studio and not just my television show?* I like to take on a challenge … many challenges, sometimes against my better judgment. I'm attracted to testing the waters of responsibility. I also like to help others, so I thought about how owning the studio would benefit me and others.

For "C," I am consistently a cheerleader for people. What I believe makes *The Donna Drake Show* "U"-nique is that, unlike many talk shows or entertainment daytime shows, I don't care to interview celebrities about their riches or who they're currently dating. I want to know what their "it factor" is, the fire that fueled them to pursue a dream that once may have seemed intangible. I want my audience to see how every person I interview is helping others and sharing their unique gift with the community.

By owning this studio, I would be able to "S"-hare my resources with artists and creators, helping make their dreams come true by producing the highest quality product. To share what I have learned while wearing many hats in the industry and no longer keeping the strengths I have used to grow my show to myself.

This thought process pushed me to take the next step in acquiring the old CBS, WLNY studio. It helped me gain the nudge I needed to believe in myself, so I took out a business loan from the SBA. I wrote Michael Pascucchi, the man who owns the building, a heartfelt thank you note that said, "Dear Michael, I would like to come home." When he received that letter, he rented me the space and showed his gratitude by paying to install a brand new COVID-19 filtration system, as well as brand new carpeting, and a fresh coat of paint. And that's where we began.

Now in the second year of building my own creative space, the physical building and networking opportunities and lifelong partnerships have expanded. We added a video podcast room complete with a green screen wall and a selection of nearly 200 virtual backgrounds, shot with three cameras in 4K quality. In this new space, I stay true to the "C" in my F.O.C.U.S., consistently being a cheerleader for other people, renting the studio to help make their dreams come true by creating their own shows. For instance, my friend, Harlan Friedman, can express his passion for street fashion while Rich "Big Daddy" Salgado has pivoted from insurance for professional athletes into sports anchoring.

I promote our shows and many others through my recent partnership with distribution streaming services, such as BINGE Network and VideoElephant which reach households worldwide. Through this, I have been able to obtain national distribution through CBS affiliate channels in major cities across the United States. I now have the privilege of saying "check your local listing" in every promotional poster. I aim to constantly grow and improve myself, investing in new technology to advance my craft. I'm very excited about a new project—a new cycloramic studio space which will project whatever fantastical, interactive setting one can imagine with touch screen elements. The possibilities are endless at Drake Media Studios!

Now that you have the means to find your strengths, it's time to stop holding off on your dream and make it happen! Drake Media Studios would not have grown and created a life of its own without my fantastic crew (who helps film my show each month as well as other projects), my interns, my family, and of course, the creative artists who make the studio their second home.

Still, a venture like this or something else all starts with you. Whether it be something small, such as speaking in front of your colleagues for a presentation or something much bigger like quitting your job, or starting your own business, F.O.C.U.S. (Figuring Out Consistent Unique Strengths) is a new tool in your mental toolbox. It allows you to gain the confidence to embark on your own venture to achieve your own aspirations. *What will F.O.C.U.S. do for you?*

Cornelia (Connie) Sexauer, dear Godmother, thank you for the opportunity to be included in Marla's books. Forever in my heart and now forever in print together. – Donna

About Donna

Donna is the owner of a certified and qualified full service production studio providing services for clients coast to coast. Recent clients: Discovery, Walmart, CBS, Novartis, and more! In addition to the video clients, Drake has her own CBS talk show, is a two-time Telly Award winner, now in its 17th season, focusing on topics of hope, motivation and resilience, through empowering conversations with celebrity entertainers, sports legends, top journalists, business moguls, authors, motivational speakers, medical professionals, and everyday heroes–sharing their stories of triumph, inspiration, and perseverance. Drake is a single mother of three adult children. Drake is known for her work as an international award-winning creative artist, writer, producer, TV personality, and SAG actor. Drake recently appeared in three feature films as herself. She is a Goddess Artemis award winner, and she was awarded the Global Citizenship award from the United Nations.

The Fruition of Donna's Studio

December 2020 - Future audio booth and edit rooms.

Old WLNY newsroom where Donna worked in 1997, which would become the new "The Donna Drake Show" set.

This Oprah Winfrey poster that hung on the WLNY TV station wall, now hangs in one of Donna's edit rooms, still inspires me daily.

January 2021 - The set for Donna's studio is complete!

January 2022 - New green room is created.

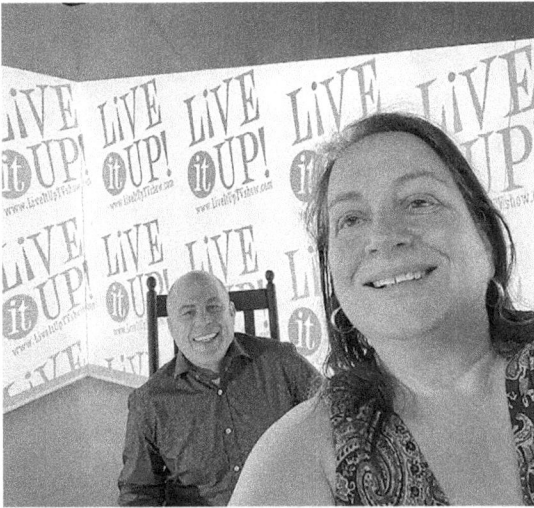

August 2022: Progress on Donna's new interactive studio, with a cycloramic wall.

New Syndication in Top DMA Markets April, 2024.

Set 2024 with the NYC skyline removed to reflect being broadcast in multiple cities. Guest: Rich Funk of BOOM CHAGA.

PETER MARKS

"There are only two ways to live life. One is as though nothing is a miracle. The other is as though everything is a miracle."

– Albert Einstein

The Gift of Dyslexia

Believe That You Are Perfect in the Eyes of Our Creator

Elementary School brought challenges for me. I could not concentrate on what the teachers were saying, plus I had difficulty relaxing in class. My behavioral problems threatened my ability to learn how to read and write. I would get teased and tormented on the school bus because of my slower learning skills and lack of coordination which I would find out later was due to my weak eye muscles.

I lived in Palisades, New York. My father was an executive salesman for women's apparel in Manhattan, and my mother was an administrative executive with the Lamont-Doherty Earth Observatory in Palisades. At that time, it was known as Lamont Geological Observatory. The observatory is a division of Columbia University, and my mother worked with scientists who studied and engineered submarines.

I started third grade at the private Rudolf Steiner school known as the Green Meadows School in Suffern, New York. Rudolf Steiner, born in Austria in 1861, was a social reformer. By the end of the 19th century, he gained recognition as a literary critic and published works, including *The Philosophy of Freedom.* His 1907 essay, "The Education of the Child," describes the significant phases of child development that formed the foundation of his approach to education. He believed that school environments should thrive on creating a social, cultural, and learning environment that recognizes the child's spiritual freedom and growth.

While attending third grade, I struggled. The teachers continued to think I was slow, mentally disabled, and impaired, so I was placed in special education. The school assigned me a paraprofessional which is a person who sits with the student and monitors what they are doing. They ensure the child is paying attention to the teacher and they understand what the teacher is saying. The paraprofessional also monitors what the student writes down on paper to ensure he or sure understands the assignments and homework.

I felt isolated and different, and I suffered from depression. I was disobedient and defiant. Kids were relentless; they bullied and made fun of me. They would call me names, push me to the ground, and beat me up when we would get off the school bus. I would come home from school, and I didn't want to do anything. I wouldn't do my homework, and I became aggressive. Finally, my mother contacted the school. She was worried about my poor grades and my bad behavior.

The school told my mother that I would have to leave unless she and my dad sought extra help for me. The school suggested that my IQ be tested. My mother made an appointment with David Wechsler, a psychologist in New York City. Wechsler was a Romanian-American psychologist. He developed well-known intelligence scales, such as the Wechsler Adult Intelligence Scale (WAIS) and the Wechsler Intelligence Scale for Children (WISC). A Review of General Psychology survey, published in 2002, ranked Wechsler as the 51st most cited psychologist of the 20th century. I went through extensive testing, and it was discovered that I had severe dyslexia.

Dyslexia, also called a reading disability, is a learning disorder that involves difficulty reading due to problems identifying speech sounds and learning how they relate to letters and words. Dyslexia affects the areas of the brain that processes language. People with dyslexia have normal intelligence and can succeed in school with tutoring or a specialized education program. Emotional support also plays an important role.

Although most people with dyslexia have normal vision, I did not. I had double vision because my eye muscles were weak and undeveloped. I had to wear large, ugly prism glasses to see correctly. I had problems with not only reading but with transcribing. I also lacked coordination due to my double vision which challenged my ability to play sports in my first two years of school.

My parents thought about enrolling me into a technical school to learn a trade, but David Wechsler encouraged them to keep me in Rudolf Steiner because I had a very high IQ. Wechsler told my parents to be patient because I was not intellectually impaired even though I needed help. He explained that dyslexia is a learning disability. It is a neurological disorder and has nothing to do with a person's intelligence. We later found out that Albert Einstein did not learn how to read until he was nine years old and that Charles Darwin and Thomas Edison both struggled in school.

I entered extensive therapy with a psychiatrist named Dr. Margaret Lawrence. I learned how to read better and started to control my anger through anger management. I also learned socialization skills even though my classmates made fun of my prism glasses. Dr. Lawrence introduced me to the world of sports. She thought sports would give me the confidence and self-esteem that I desperately needed. I started to play baseball and basketball both at home and school.

My learning disability had a significant impact on my family environment. My sister, Nan, felt neglected and jealous I was getting all the attention. She developed behavioral problems when she was 12 years old. She went to a private catholic school, and at night she and her friends would disrupt the nuns living there by lighting firecrackers.

By the time I entered junior high, Palisades had become a rough neighborhood. When my parents realized there was a drug infiltration in Palisades and Nan was involved with drugs, they decided to move to Teaneck, New Jersey. Eventually, Nan attended a private school in New Jersey called Dwight School for Girls. The school developed her writing skills and prepared her for college at Ramapo College of New Jersey and Columbia. Nan graduated with top honors from Columbia in 1975 and became a licensed social worker.

I entered Thomas Jefferson Junior High in Teaneck, a mainstream school system. I fought my way through junior high and made up three years of education in one year. I would lock myself in my bedroom during the weekends and read any book I could get my hands on. I self-taught myself, and my vocabulary improved.

In junior high, I was approached by the biggest bully in the neighborhood. His name was Victor. He and a few of his friends started to bully me. Victor put his hands on my shirt and pulled me up by my collar. He was getting ready to push me to the ground. After being consistently bullied, verbally and physically, since first grade, sheer anger and adrenaline poured out of me. I know that physical violence certainly shouldn't be one's first response to bullying, but sometimes it's necessary. If a bully is beating you up at the present moment, I believe in ditching all consideration of getting in trouble and using any means to get out of there. I knew I could not make Victor stop; I had to retaliate in a physical way. I do not know how I did it, but I returned his push with a stronger push—his punch with a more forceful punch. Let's just say I finally evened the score. By this time, people were coming out of their homes to watch the fight. My mother tried to stop the fight, but my father stopped her. He told her I needed to prove myself.

I did not care if my parents blamed me for the fight, and I did not care what people thought of me as they watched me fight. I had two options: I could continue being bullied and sink into depression, or kick him against the curb once and for all. This fight was not sinking to the bully's level but rising above it and taking a stand for myself. He was in bad shape after the fight. I was never bullied again.

I graduated from Teaneck High School with a B plus average. My parents and Dr. Lawrence encouraged me to go to college. I graduated from Ramapo College of New Jersey with a Bachelor of Arts in Sociology and a minor in business. I continued my education and received a certificate from Cornell University in psychology dealing with children on the autistic spectrum.

When I look back on my school years, I have realized that my parents made sacrifices and worked hard to understand my disability. I am sure they were overwhelmed while working with me and did their best to encourage me. They were always concerned about my emotional and mental development, and they allowed me to learn things at my own pace. Whether I started a new school, had a new teacher, or joined a new sports team, my parents advocated for me every time I needed special help. They always got their point across in a friendly but clear and firm manner. They provided me with professional counseling support at an early age that was vital to my learning.

After graduation, I met a man who would change my life. Upon talking, we both realized we had a desire to make documentaries. Soon after our first discussion we put our words into action and contacted the local East Orange, New Jersey, Cable Television Company to sign out some cameras. Our first documentary, *Thievery in Suberia; You Can Fight Back, in Ho-Ho-Kus, New Jersey*, won an award. We followed the sheriff of Ho-Ho-Kus around and went on emergency calls and interviewed people who were burglarized and victims of other crimes.

I then started the cable television show called *You, The Consumer.* I went to local businesses and charged them $700 for promotion on my cable show. I contacted doctors, charged them each $1,200 to $2,000 and gave them five TV airtime advertising spaces in the studio where I worked on network cable.

I eventually started my own business as a documentary film producer and began *Medical Magazine* for East Orange, New Jersey Cable Television Company. The program focused on health and medical issues comparing prescription medications and surgical treatments. I then developed a travel show called *America Coast to Coast.* I highlighted places like Martha's Vineyard, Nantucket, and Novia Scotia. I traveled the East Coast and explored different bed-and-breakfasts and privately-owned motels. I also created *Today's Gourmet* with Bob Lape. Lape, a broadcast journalist, writer, restaurant reviewer, and food critic, inspired my interest in food. We would travel to restaurants in New York City, other local areas, and the East Coast. Bob would sample the food, analyze its presentation, rate the service, and describe the restaurant's ambiance. He would end the show by reviewing his full dining experience. Later, I won producer of the year for East Orange New Jersey Cable Television.

My last show, *Matters of Life and Death*, was with Dr. Earl Grollman. Dr. Grollman was a Rabbi and a pioneer in the field of death and dying. He was widely known nationally as an expert in grief counseling. Grollman headed up the Beth El Temple Center in Belmont, Massachusetts, for 36 years before retiring. His books included *Living When a Loved One Has Died* and *Straight Talk About Death for Teenagers: How to Cope with Losing Someone You Love*.

Several years later, I was diagnosed with Epstein Barr virus. I became very sick and had no energy. I was not able to work for quite a while. My Aunt Joan, an astrologer, suggested that I study Astrology until I was able to return to work. I knew that I was intuitive and psychic since I was 11 years old. I gave psychic readings at the cable TV studio. I attended a correspondence school at the Wright Institute for Astrological Studies in Canton, Ohio.

Once certified as an astrologer, I started doing readings for people locally and attended psychic fairs. My work and I became a household name, and I was able to meet Janet Russell, a well-known psychic medium in Medford, New York. I appeared on her TV show *Beyond the Unexplained* and began to meet independent psychics and appeared with them in their separate show segments as well.

I presently have my own YouTube channel called *Readings by Peter Marks*. I am a spiritual advisor, psychic medium, astrologer, media personality, and author. I read people from all over the world and from all walks of life. I try to provide insights on career, family, business, and personal issues in a warm and friendly style. I believe that faith is a powerful force that can shape lives and society in positive and productive ways.

I recently formed a strategic business alliance with one of the most famous media marketing experts in the United States. His name is Stefan Rybak, and we created Marks Rybak Global Media. The company produces exclusive long-form and short-form video and audio programming content that informs, inspires, and entertains viewers and listeners and provides guidance, direction, and enlightenment. Production includes, but is not limited to, streaming, media, books, and a multimedia platform that offers resources on education, encouragement, entertainment, and self-improvement to a global audience.

Marks Rybak Global Media's podcast, *Find My Divine,* is televised throughout the United States, Great Britain, Scotland, and Wales. On March 22, 2022, my first book, *The Gourmet Cookbook For Astrology Lovers,* was published.

As I have lived my life, I have learned an important lesson. I try to teach this lesson to my clients, parapsychology students, friends, and foe. At heart, we all have the capacity to heal ourselves and nurture others. In the long run, I have also realized there is a much higher power that exists and controls the universe. We each have the ability to tap into this higher power to overcome our obstacles and achieve great success. And that's how I took back my power.

References
1. Frank, George (1983). *The Wechsler Enterprise: An Assessment of the Development, Structure, and Use of the Wechsler Tests of Intelligence.* Oxford: Pergamon. ISBN 978-0-08-027973
2. Grollman, E. *"Living When A Loved One Has Died."* Boston, MA: Beacon Press, 1993.
3. Grollman, E. *"Straight Talk About Death With Teenagers: How to*
4. *Cope With Losing Someone You Love."* Boston, MA: Beacon Press, 1993.
5. Isaacson, W. *"Einstein: His Life and Universe."* NY: Simon & Schuster, 2023.
6. Johnson, P. *"Darwin: Portrait of a Genius."* NY: Penguin Books, 2013.
7. Marks, P. *"The Gourmet Cookbook For Astrology Lovers."* Mukwonago, WI: Nico 11 Publishing & Design, 2022.
8. Morris, E. *"Edison."* NY: Random House Trade, 2020.
9. Steiner, Rudolf. *"The Philosophy of Freedom."* Independently Published. 2018.
10. Steiner, Rudolf. *"The Education of the Child."* Rudolf Steiner Press, 1965.
11. *The International Board of Directors. The International Dyslexia Association,* 2002.
12. Wechsler, D. *Wechsler Intelligence Scale for Children, Fourth Edition*
13. *(WISC-IV). APA PsychTests,* 2003.
14. Wechsler, D. *Wechsler Adult Intelligence Scale, Fourth Edition, (WAIS-IV).* APA PsychTests, 2008.

About Peter

Peter Marks was born in New York City and raised in Palisades, New York. He received a Bachelor of Arts in Sociology and Business from Ramapo College, New Jersey. In 1993, Peter attended the Wright Institute of Astrological Studies and graduated with honors.

From 1979 to the early 1990s, Peter was the producer of many television shows. One of the most notable was working with Bob Lape on *Today's Gourmet, Bob Lape*. His association with Bob Lape, a food critic and award-winning journalist inspired his interest in food. Peter later worked with Bob on a travel and food show called *America: Coast to Coast*. As Peter worked with Bob, he developed a taste for some rare and exotic dishes. Bob Lape gave Peter the experience of a lifetime, as he taught him the alchemy of cooking and food presentation.

Peter has been a practicing astrologer for over 20 years and has his own YouTube channel where he speaks to clients in person and over the phone. He is also featured with Janet Russell on the TV program *Beyond the Unexplained* in Medford, New York, and with Joyce St. Germaine, host of the TV program *The Sacred Journey* in Farmington, Connecticut. Peter also does astrology readings at The Connecticut Sanctuary located in Waterbury, and he has a private clientele that has expanded over the years.

Peter has also appeared at The Society Room in Harvard, Connecticut. He has conducted readings at Mar-a-lago in Florida, and has done readings for many of President Trump's friends. Peter has given his astrology readings to members of The Royal Family.

He was featured in *The Litchfield County Times, The Huffington Post*, and the *Jim Masters TV Podcast*, JoAnna Garfi-McNally podcast called *The Soul Searcher*, and *I've Got Your Number*, a blog talk radio show with Lois Martin and JoAnn Garfi-McNally.

He is also doing a successful weekly podcast with Theresa Connolly Seichter and Ashley Banta Cosmicteahour 999.

Peter is looking forward to meeting and working with people in the New Age.

For more information, please visit peter-marks.com

Peter at 8 years old.

Peter and his sister, Nan.

Peter's parents on their anniversary.

Peter celebrating his sister, Nan's wedding.

Peter and his mom, dad, Nan, and Jack, at his sister's wedding in Ramsey, New Jersey.

Peter with actress Angie Dickinson and her manager on a cruise ship.

Peter and his grandmother celebrating his birthday at his estate in Connecticut.

Peter at Mar-a-Lago.

Peter receiving the Cape Award for Producer of the Year for TV excellence in programming on public TV, 1988.

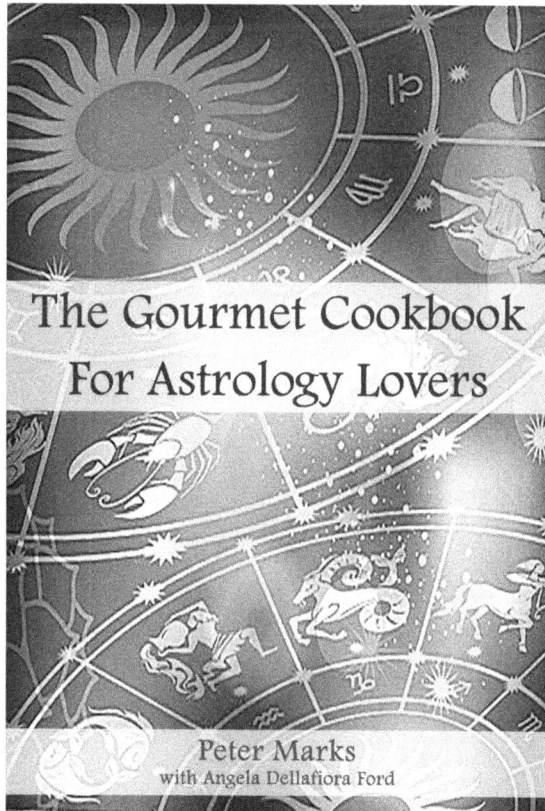

Peter's cookbook.

LAURIE RAÉ GRAHAM BENNETT

"I must be a mermaid ... I have no fear of depths and a great fear of shallow living."

– Anais Nin

It's All About Karma
God's Blessings

I'll never forget that horrible day in March 1981. My parents sat me down at the kitchen table and told me I'd need a third open heart surgery. I turned 13 years old earlier that month.

I was born with four separate heart defects: structural malformations or missing parts. The most serious of my four defects is a missing ventricle. The two ventricles are the "pumps" of the heart, the part of the heart muscle that contracts and releases to send blood to the lungs and the body, then back to the heart. With only half a pump, I have immensely reduced activity, energy, and oxygen levels compared to a heart-healthy person of the same age. There is no fix, no correction, and no cure for my defects.

Desperate for answers upon the diagnosis I got at three weeks old, my parents would simply not accept the prognosis given them. Doctors told them I had six months to live, and then said to "take me home to die." Instead, my mom and dad appealed to Texas Children's Hospital for a consultation with the team of Dr. Denton Cooley. He is considered the world's greatest heart surgeon, often working on children whose hearts were the size of walnuts. He agreed to help me, performing first one, and then another, open-heart surgery on me. Those two surgeries, at ages 11 months and 35 months, were to keep me alive long enough for the next medical advancement in the very young field of pediatric cardiac surgery. The plan worked.

By age 12, however, I was declining quickly. I was so weak that I could barely make it from class to class while at school. I had a bus that picked me up at my front door and dropped me off directly in front of the school. I couldn't carry books, so each classroom had a copy for me, and I had a stack at home. I even kept my brown bag lunch at the school secretary's desk. I felt exhausted all the time.

That awful day in our home, my parents explained what they knew. My local pediatric cardiologist had heard of a new surgery called the Fontan, and he had already told my parents I was a perfect candidate for it. The doctors still considered the surgery experimental and only two surgeons in the country were capable and willing to do it. It would take 14 hours, and it was possible I would not survive. Expecting it to be painful, and feeling very scared, I told my parents I didn't want to do it. Genuinely, I suggested perhaps it was my time to die. My parents overruled me. I was powerless over their decision, and frankly, over the entire situation. I was terrified.

I told none of my friends, choosing to keep the news private. My family was told, but most everyone didn't understand my defects, let alone the basics of the complicated surgery I was facing.

When we met with the doctor, he asked me what I most wanted to physically be able to do. I told him I wanted to roller skate with my friends in my neighborhood, and I wanted to ride my bicycle. Based on successes he had already had, the doctor replied, "If you make it off the table … I promise you'll be roller skating in six to eight weeks." I reluctantly agreed to have the surgery, hanging my agreement on his promise. He began discussing a date right away. I told him we weren't doing anything until after June, when school ended for the year. After all, I was a straight-A student. He told me I wouldn't make it to June, and like a cornered, powerless animal, I curtly barked at him, "Watch me!" Despite my bravado, I finished seventh grade thinking I might die.

Luckily, it was also March when I first heard Rick Springfield's song "Jessie's Girl" on the radio. I was instantly hooked, and I bought his album *Working Class Dog* right away. To comfort myself while waiting for the surgery that would either change—or end—my life, I played the album over and over. I was in love!

At the end of June, my parents and I traveled from our home in Los Angeles, California to the Mayo Clinic in Rochester, Minnesota. No one prepared me for all I would face there. I wasn't told how badly it would hurt; I wasn't told about the pre-op procedures; I wasn't even told I'd have a scar! No one knew how my body would react to the drastic changes made inside my heart and to the redirected flow of the blood to and from my heart and lungs. No one predicted the multiple daily blood tests I'd get that made me phobic about needles or the months afterward I'd spend vomiting every day from the physical trauma of the surgery. No one even considered the emotional toll it would take on everyone in my family, especially me. And no one expected this could change my life for the worse.

Instead of roller skating in six to eight weeks as promised, I was bedridden for over a year. After 18 months, I had only reached the same energy level I had before the surgery. The doctors and my parents had no answers and could not improve or fix my situation. The medical team told us to just give it time. *How much more time was I going to have to suffer?*

Too weak to attend school with my friends, I was lonely and inconsolable. I missed my entire eighth grade year. Every day I took a total of 90 pills, just below toxic level doses. My health status was incredibly perilous. I became deeply depressed and shut out everyone, even my parents. I felt furious—at God, my parents, and the doctors—for all of my physical and emotional pain. I wished I had died—could die … but couldn't figure out how to make it happen. I refused to discuss my experiences with anybody, not even my family. When a psychotherapist came to our home, I sat in silence with my arms crossed for the entire hour. There was no way she could understand; *What was the point of talking?*

However, for 45 minutes every day, I embraced the single joy I felt when I played Rick Springfield's breakthrough album, *Working Class Dog*, on my dad's record player. Rick was there for me at the most painful time in my life, every day, singing his heart out about emotions I was too fragile and scared to express myself. Each day when I moved the record player needle over, Rick and his music gave me hope and courage. Somehow, I felt he understood my pain. As a magnificent Grammy award-winning songwriter, Rick is criminally underrated ("In Defense of Rick Springfield" by Dave Lake, March 17, 2015) and searingly candid (press release for *Late, Late at Night*, Rick Springfield's autobiography, Dec 14, 2009).

While the situations he sang about were nothing I had experienced, the feelings were relatable—pain, abandonment, rejection, fear, loneliness, and depression. Of course, he was handsome too, and had a smile and a voice that gave me butterflies. But my biggest connection was to his lyrics. His music wasn't "Don't Worry, Be Happy." He wrote about his own struggles, his emotions about universal issues. Rick's music helped me to survive, then pushed me forward into healing.

The physical scars faded somewhat, and my health and energy improved over the years. By then, Rick's music (including 12 more albums) was my constant companion. His music had spoken to me, reached me, in a way no one else ever could. In both the tough times and good times of my life, I've always come back to his music as my emotional release.

Twenty-four years after my surgery, in May of 2005, I found an opportunity to meet Rick Springfield in Tucson, Arizona. Told I'd get about two minutes to talk to him, I was a mess trying to figure out ahead of time what to say.

My husband, Jay, helped me simplify my overcrowded thoughts by asking me, "In a few words, what do you want to tell him?"

Immediately I replied, "Thank you."

"Then write him a thank you note," Jay responded.

So, I did. I wanted to thank him for creating the music that sustained me in the darkest time of my life. Since I figured my nerves would probably interfere with my ability to speak my message clearly, I wrote out exactly what I wanted to say. I spent practically 10 sleepless nights writing and rewriting every single word until my "thank you note" was perfect; plus, I practiced reading it in the allotted time. My sweet husband agreed to fund the out-of-state trip to make this magical experience happen for me and as a great excuse to celebrate our eighth wedding anniversary.

When the day arrived and Rick walked into the room, I froze. Jay nudged me toward Rick, and I asked if it was okay to read my letter to him. This was my moment, two and half decades in the making. With no expectations, and courage I didn't foresee, I told Rick what I had been through all those years ago. I explained how physically ill I had been, how I had completely shut down emotionally, and how his music was the only brightness I experienced while I fought to stay alive. It was the first time I had opened up to anyone except Jay. He listened with rapt attention, never once taking his eyes off me. He moved closer and closer to me, listening to every word I spoke. All the raw feelings from those painful years rose to the surface with a force I didn't expect. I sobbed. Rick put his arm around me and held me tight. I noticed he, too, had tears in his eyes. He gave me a sincere, warm hug and thanked me for sharing my story with him. He was incredibly compassionate and willing to give me the time I needed to share my story with him. I collected my composure enough to get a photo and asked him to sign my original *Working Class Dog* record album, stating, "This is where it all started." Rick wrote my name on the album cover, paused, and then wrote, "Glad I was there! With love, Rick Springfield." I truly feel his soul heard mine in that moment. We had made a true and deep connection. And it changed my life.

The concert afterward was a blast, and I celebrated in all the ways my 13-year-old self wasn't able to do. I sang along at the top of my lungs and even danced a little. I managed to get on stage with a large group of fans when he sang "I Get Excited"—a concert tradition at that time. Then Rick picked me to sing the song with him, holding me between him and his guitar! I don't know how I didn't faint, but I guess I did get awfully pale. Then, the encore song was the one I've had "pinned on my heart" since the very beginning—"Love is Alright, Tonite." Unbeknownst to me, Jay had told him at the meet and greet it was my favorite song. Rick sang an entire verse to me. At the end of the song, Rick bent down in front of me, then obviously and deliberately handed me the guitar pick he used for that song.

Afterward, back at our hotel, I couldn't fall asleep for hours. Finally, I collapsed from the excitement of meeting Rick and dancing and screaming at the concert.

When the alarm rang the next morning, I excitedly told my husband, "It's gone, it's gone!" Jay, still not fully awake, asked me what I meant. "My anger is gone," I clarified. I physically felt lighter. I had finally let go. The anger from my past experience that engulfed me since I was a young teenager was gone. Rick's genuine, caring reaction to my story gave me the freedom to feel safe to open up regarding my heart problems. I had lived 24 years with this undeniable anger, but it only took one afternoon, and a magical—no, it wasn't magical, it was spiritual—a spiritual meeting with my favorite musician and songwriter, to find acceptance. My story had come full circle. I received validation and reassurance from the only person able to help me when I desperately needed support.

The weekend had far exceeded anything I could have ever imagined in my wildest dreams. I thought nothing could have made it any more perfect. But there was more to come!

When we arrived at the airport to fly home, a TSA agent pointed to the Rick Springfield T-shirt I wore and said, "I just saw him about 10 minutes ago." I couldn't believe he was so close! Spotting Rick far ahead, Jay got caught in the TSA line. He told me to go ahead without him. I walked as fast as my heart would let me, and I found Rick in the gift shop. Not shy this time, I boldly went over to say hello. He asked if I had enjoyed myself the day before and I said, "Of course! It was the best day ever!" I asked him what flight he was on—hoping he was on mine—and he replied with frustration that his flight had been delayed four hours already.

Out of my mouth came the words, "Rick, you can have our tickets." They literally just popped out, no forethought, no scheme, no plan, nothing. He graciously replied, "No, that wouldn't be right." I repeated my offer. Again, he politely thanked me but declined. I wished him well, hugged him, and said goodbye.

A few minutes later, I saw his manager at the gate counter, trying to find out the status of their flight. Since I had met him the day before, I repeated to him what I told Rick about giving up our seats. He quietly told me Rick had a funeral to attend (later, I would find out it was for a woman who had been like his second mother for over 30 years and the mother of his lifelong best friend). He said that the airlines probably wouldn't allow me to switch the tickets anyway. Then he walked away.

I felt sick inside. I wanted to do something, anything, to give back to Rick, to the man who had given me so much for so long. Even though I wasn't close to God, I said a silent prayer in the hopes I could help. Jay agreed I should try again. Then I approached the Southwest Airlines counter. I asked if I could exchange our two tickets for the delayed flight so my friend could get on this one and get home in time. In all her years with the airline, the agent told me that asking to go on a delayed flight was the strangest request she had ever heard! I desperately tried to convince her how serious I was, promising not to complain about whatever our wait would end up being. After talking to her supervisor, she agreed!

I raced back to the gate where Rick was sitting with his manager and band. Rick was on the floor, in the corner. He had his knees bent to his chest, all hunched over as he tried hard to listen to whomever he was talking to on his cell phone. Then he saw me standing there and looked up at me with a face full of frustration and stress, likely not in the mood for a fan interaction. I said, "I really don't mean to be a pest, but the airline said it was okay. You can have our tickets."

He said, "Really?" His face completely softened. I noticed his eyes tearing up. He smiled only a little, as if he didn't believe me.

I repeated, "Yes!"

"That is so awesome," he replied.

He stood up, told his manager to call his driver, and the three of us walked over to the ticket counter to do the paperwork. I motioned to my husband Jay to come over, since he was holding both of our IDs and boarding passes. The agent processed Jay's ticket first, exchanging his ticket for Rick's. Then she asked, "Who am I exchanging Laurie's ticket for?"

His manager said to Rick, "Well, who do you want to go with you, me, or one of the guys?"

Rick paused, looked at his manager, looked at me and then looked back to his manager. Rick sternly but nicely said, "No, we're only exchanging one ticket. I'm flying home with Laurie!" Looking at Jay, Rick offered his hand and continued, "If it's okay with her husband."

My amazing and loving husband replied, "Of course!" shaking Rick's hand.

My mind couldn't focus. I told Rick, "I have to sit somewhere," and he replied, "Sweetie, you're sitting with me." That's when the weight of his gesture sank in. I couldn't believe that now, in his own moment of grief and despair, Rick had given to me yet again. I was stunned and deeply touched.

"So, you don't think I'm a wacky fan for chasing you down in the airport?" I asked Rick.

He replied, "Yes, but you're a good wacky! You gave, it wasn't take, take, take. What you did was the best thing a fan has ever done for me."

By now the gate agent had recognized Rick and offered to pre-board him. I asked if I could go with him, but she replied, only if you are disabled. Well, I sure the heck was and for the first time, I was proud to proclaim it! I said, "I have a heart problem" before she could finish her statement, and she gladly gave me the pre-boarding pass. There was no way I was going to miss out on even one minute of time with Rick.

We were called to board right away. I said goodbye to Jay, kissed him, and in-between my tears whispered, "I love you."

He smiled at me, and as I walked down the jetway with Rick, Jay said, "Happy Anniversary!" If there is a husband of the decade award, he just won!

I got on the plane with Rick Springfield, and I had an entire hour and 15 minutes to talk with the man who had been my idol, my rock, my unfailing inspiration, and joy for a quarter of a century.

Sadly, the weekend of a lifetime came to a close, yet I was still on cloud 10. But I couldn't stand not knowing if Rick had made it on time that day to the service. Two weeks later at a local concert, I went backstage again. Rick noticed me right away, and his face had a look of sudden happiness and surprise. He walked straight to me with open arms and scooped me up off the floor, spinning me around in an enormous hug. When he pulled back, he told me he owed me so much. *Rick Springfield owed ME?* I immediately questioned him because I felt I owed him my entire recovery!

I asked him if everything turned out okay and he said "Yes, the timing was perfect. I walked into the church right when the organist played the first notes of the first song. There was no way I would have made it on the other flight."

Next, Rick autographed the guitar I had purchased earlier. Imagine my delight when I saw he remembered not just my name but how to spell it. I watched him write "To Laurie," and after he paused, he wrote "My Airplane Girl," drew a heart and signed his name. I couldn't believe it—he gave me a nickname!

If I had written the entire experience as a movie, I couldn't have written it to be as extraordinary as the reality actually was. Moments of this magnitude don't just happen; I know now that God truly orchestrated the entire experience. I believe God put Rick into my life at the exact time I needed him, knowing his music would end up being so valuable to me. I believe God was with me during my surgery and recovery. I believe God wanted me to heal and let go of my anger I carried with me for 24 years. But I was stubborn, really stubborn. I like to say God finally had to throw a brick at me—labeled Rick Springfield—to make it actually happen. I also believe God put me in Rick's life right when he needed me to get home for such an important event.

Afterwards, I wanted to tell everyone I knew what had happened. I was so excited that Jay joked I'd stop people on the street to tell them about Rick's kindness and compassion toward me. But I couldn't just tell the Rick part of the story—in order to convey the weight of meeting him properly, I had to explain why I loved him as much as I did. This wasn't just a typical teenage crush on a cute rock star. His music was lifesaving. In order to share that part, it meant opening up about what I had been through as a teenager. The two pieces were intertwined, and both halves strengthened each other. They didn't exist without each other.

Suddenly I was telling everyone who would listen, climbing out of the hole of darkness that I held myself in for so long and taking back my power over my story—my life. Little by little, with each telling of my new story, I healed a bit more. No longer was my surgery something done to me, but an integral part of my history, and part of God's plan for my life. The bottle had been uncorked, and I sought therapy to deal with all the feelings I had stuffed and swallowed previously, and to move past the darkest moments I had experienced. I found acceptance in myself and forgiveness for my parents, the doctors, and especially God. I started asking my parents questions for clarification and knowledge. Wanting the exact date of my surgery so I could celebrate my "heart-aversary," I called the Mayo Clinic to get it. I attended a national conference held by the Adult Congenital Heart Association. Jay has told me he's never tired of hearing me tell my story. As he watched from the sidelines, he witnessed my radical change and saw his wife light up with joy.

I find no irony that Rick titled the album before I met him *Karma* and named the album when I met him *Shock/Denial/Anger/Acceptance*. Everything that could have gone right, did. All the pieces of my last 24 years fit perfectly together with meeting Rick to create a beautiful picture of healing. Jay was incredibly supportive and never even paused when I said I wanted to give up our airplane tickets. I felt incredibly grateful to have finally told Rick what he meant to me in the course of my life and, most importantly, to say thank you. I found Rick to be a kind, generous spirit, with a deep soul who wears his big heart on his sleeve. The bond that we built in those brief moments was beyond a normal fan experience and will always be one of my most treasured lifetime memories.

In the years since my transformation, I have grown in ways I could have never imagined. I feel at peace with having a rare and serious heart condition, and I've never felt happier. Now, I can talk about the traumas I've experienced. I have a renewed energy, a "new spark" as my mom calls it. I feel lighter, unburdened. I smile more. I have the courage and eagerness to confront my past. I share my story with younger heart patients and their parents, so they may know someone has been there before them, and they are not alone. While I have met other adults with four or five defects, I've never met another person with my same defects. That used to make me feel alone; but now I feel unique. I also know God never once left my side.

I used to define myself through my health history. I used to be "Laurie, the one with the heart problem." Now, I am simply, and fully, Laurie.

> *"Every little bit of love,*
> *I give to another,*
> *you know what I believe,*
> *it comes back to me"*

Lyrics from the song "Karma," by Rick Springfield.

References

"Article in Defense of Rick Springfield" by Dave Lake, March 15, 2015.

Press release for Late, Late at Night, Rick Springfield's autobiography, December 14, 2009.

About Laurie

Laurie was born and raised in Southern California and has lived there most of her life. She grew up in an average middle-class family with two parents, an older sister, and a pet tortoise. But she had a tough beginning. Born with four heart defects, she wasn't expected to live more than six months. Defying the odds against her, she turned 56 years old this year. She had two open-heart surgeries as an infant, and a third experimental heart surgery at the age of 13. Laurie attributes her life to strong, determined, informed parents, the skill of the world's most famous heart surgeon, Denton Cooley M.D., God, and her own internal "light and fight." While she survived the surgery as a teenager, many things went wrong during her recovery, and she spent 18 months in bed, missing her entire 8th grade year. In addition, the following year her parents separated, and eventually divorced. Unable to compete with her peers on the playground, she excelled in the classroom instead. She graduated in the top 10 percent of her high school class. Laurie earned her bachelor's degree in liberal arts, concentrating in technical and creative writing, from California State University, Long Beach. She earned Magna Cum Laude and Phi Beta Kappa. Due to her disability, Laurie was only able to work a few years, notably as a national Customer Service Representative at Toyota Motor Sales, USA. She also runs her own organizing business, helping "average moms and dads" get a handle on their clutter. Her favorite items to organize are photographs and paperwork. She married the love of her life, Jay, 27 years ago, and they created a home together in the local area. She is also a proud cat mom to three rescues, her "boys." Laurie enjoys spending time scrapbooking, attending Rick Springfield concerts, and doing the inner work to learn and grow as a person. Her goal in life is to be the best Laurie she can be.

Rick Springfield and his Airplane Girl.

Laurie and her two favorite men - her husband Jay and Rick Springfield.

Laurie's happy place is either at a live concert or at the ocean!

MANETTE KOHLER

"I now see how owning our story and loving ourselves through that process is the bravest thing that we will ever do."

– Brené Brown

Is That a Zebra I Hear?

As I worked at my computer on a cold and gray, Wisconsin, January day, my buzzing cell phone jolted my attention. Glancing down, I read "Racine County Health Department" across the top of my cell phone screen. I'd normally let incoming calls go to voicemail when working but I couldn't stop staring at the screen as the phone continued to buzz. *Why was the health department calling me? What could they possibly want?* I tentatively reached for the phone and answered the call. The pleasant-sounding woman on the other end of the line introduced herself and then got right to the point. "I'm calling to let you know that you have Lyme Disease," she said. I felt like someone punched me in the gut, and I felt the color drain from my face as I sat there silent, letting it sink in. "Hello?" she asked, likely wondering if I was still on the line. I replied that I was, indeed, still there. She explained that Lyme disease was a Category II reportable condition in Wisconsin which meant that healthcare providers had to report positive test results to the patient's local health department within 72 hours upon recognition of a case. She wished me well, and we ended the call.

I set my phone down and burst into tears. I mean the "ugly cry" type of tears. *Why did this news strike such a punch?* I'd already received my positive Lyme test results from my doctor just the day before and was thrilled to finally have an official diagnosis that had eluded me for well over a decade. As I sat at my desk, I allowed myself to fully feel my emotions and to ponder why I was overcome with emotion. The reason was crystal clear ... I was FINALLY validated. My symptoms were real and not "all in my head" as many had suggested within the medical community. This was 2012, and I was about to embark on my healing journey, but this is not where my story begins.

Let's go back to 1993. I told you this was a long journey, but I promise to give you the abridged version. I'd just graduated from veterinary school, gotten married two years prior, and was working at a busy veterinary practice in rural Georgia, 30 minutes east of Atlanta. When we weren't working or tending to our horses, my husband and I did a lot of exploring in the Blue Ridge Mountains of northern Georgia, hiking both groomed and primitive trails in search of the multitude of beautiful waterfalls that cascade into icy, clear, mountain streams. We encountered a myriad of forest critters including snakes, bears, birds, deer, and other small animals as well as some creepy crawly things like spiders and ticks. We'd pull off the occasional tick but didn't give them a lot of thought, and I can't remember ever seeing any sort of rash after a tick bite. Lyme existed in Georgia at that time, having first been recognized in Georgia in 1987, but it wasn't prevalent like it was in the Northeast U.S., so it wasn't really on our radar. In fact, in the two years I'd worked as a veterinarian in Georgia, I'd never had a Lyme test come back positive for a dog.

I was 29 years old at that time and very healthy and fit, but one day a bizarre illness hit me like a Mack truck. I was suddenly gripped by flu-like symptoms including a high fever, terrible headaches, dizziness, brain fog, sore throat, and I lost over 15 pounds. This illness went on for over a month, and I was assured by my doctor that it was probably just some virus that would pass with time. A 10-day course of antibiotics was prescribed at one point in case I had a sinus infection but still no relief from my symptoms. As time went on, I felt and looked sicker, and clients started to comment on how tired I looked and how thin I'd gotten. Looking back, I now realize how poorly I advocated for myself. I was shy and reserved back then, but I was also in a foggy haze from the illness. Perhaps I didn't accurately portray my symptoms to my doctor or convince her just how badly I felt. I figured my doctor would know what to do; I trusted her. If only I'd pushed for an infectious disease work up, my story might have been a very different one. That illness finally did wane away, and life went on as usual.

Fast forward to 2003. I was a mother of two girls, ages three and five. My husband and I followed our roots back to the Midwest just three years prior and bought a home in southeastern Wisconsin. I worked at a veterinary clinic in our town and spent my free time enjoying my family and pets. Life was good. Until it wasn't. Odd symptoms in various body systems started popping up, beginning with dizziness, excessive tiredness, and horrible headaches which started to occur more and more frequently until there was rarely a day that I wasn't dizzy. Migraines were happening several days per week. One day, my daughters asked me if I felt well enough to play, and that memory just breaks my heart. My physical health was affecting them too. I'm not one to run to the doctor for every little thing, but it was clear that I had to see my doctor. The dizziness and headaches triggered vague memories of my "mystery illness" I'd battled in Georgia a decade before. Basic lab work came back normal, but my thyroid panel was consistent with hypothyroidism. Perhaps that's why I was so tired? I started thyroid replacement therapy and was referred to a neurologist for the dizziness and migraines. Little did I know that she was the first of a long list of "ologists" I'd see over the next several years.

It is worth noting that since returning to Wisconsin, I was routinely diagnosing Lyme and other tick-borne diseases in my canine patients, and it was becoming abundantly clear that the upper Midwest was a growing hot spot for Lyme and associated diseases. Lyme was on my radar now, at least for my patients, and it soon would be for me as well.

While I worked with my neurologist to find a cause for my migraines and dizziness, trialing one medication after another to try to keep it all at bay, other body systems started to have issues too. First it was joint pain, and not just in my knees which had always caused me issues. It was new joints swelling up … fingers, thumbs, hip. My primary care doctor sent me off to a rheumatologist for the battery of tests, all coming back within normal limits. My joints quieted down over time, but inflammation would flare up from time to time.

Then my gastrointestinal tract started acting up with debilitating abdominal pain. Off to the gastroenterologist (my third "ologist" if we're keeping track) I went. Tests revealed I had colitis and ulcers in my colon, and I began treatment for that. Meanwhile, my migraines were still raging on. She suggested that my migraines were, perhaps, just stress related. I had no doubt that stress was a factor, being that I was a busy working mother of two, but I knew it wasn't the whole story. I took up yoga, tried to eat healthy, got into an exercise routine, made time for myself, and focused on getting more sleep at night, but the migraines did not decrease in frequency.

With no answers for the migraines, and with a new symptom of brain fog which caused me to forget what I was saying, mid-sentence, or forget why I entered a room, I switched to another neurologist. I now had three body systems having issues, so I presented this new neurologist with my whole list of what must have looked like "unrelated symptoms." Or were they? Was it possible to have multiple, separate issues? At this point it was 2005, and I was only 41 years old, but I felt like I was falling apart.

With joint, central nervous system and gastrointestinal tract issues, Lyme was now on my radar. Frustrated with lack of answers, I did what most folks tend to do. I turned to the internet, something I typically dread my own clients doing because of all the questionable information out there in cyberspace. I followed the science. I sought out scientific studies from peer reviewed journals, articles written by folks with "MD" and lots of other letters behind their names. I wanted to figure out how my symptoms might all be related. After all, this is how I approached my patients' cases—like pieces of a puzzle that might somehow fit together to give me the answer I sought. I was persistent in finding answers for my clients and their pets, and now it was time to dig in my heels and be persistent getting to the root of my problems. I wanted so badly to feel good again.

My symptoms seemed to fit with several differentials including chronic fatigue syndrome, autoimmune disorders (including Rheumatoid arthritis and Multiple Sclerosis), fibromyalgia and Lyme disease. The one that seemed to fit best with my symptoms was Lyme. It was well-known for causing joint pain and arthritis. My headaches, fatigue, and brain fog also fit with Lyme along with a new symptom of neck stiffness and pain. I also learned that Lyme causes gastritis, duodenitis, and colitis. In hindsight, I wished I'd asked the gastroenterologist to submit biopsy samples for Lyme testing, but I hadn't yet discovered the Lyme-gut connection at that time.

I shared my research and ideas with the new neurologist. My expectation was that he'd have an open mind and consider my idea of one etiology for some, if not all my symptoms, including the possibility of Lyme disease. After all, something was causing inflammation throughout my body, so it made perfect sense to me. What I didn't expect was his immediate closed-off demeanor and raised eyebrows. You'd think I'd just suggested that I might have Ebola virus. He made it clear that he wanted to focus on just my migraines and felt that a Lyme test was a waste of time. Feeling defeated at that moment, I listened as he instructed me to keep a list of my foods to look for possible triggers for my migraines, and he told me to decrease my stress. Ugh! I felt unheard, plain, and simple. He did order an MRI which was normal, so he felt I did not have MS. Any time I could rule out a problem was also a win in my book.

I was frustrated that I had no answers and that I felt dismissed by the neurologist but, to be fair, he was basing his decisions on what he knew and what he'd seen in his training and career thus far. Despite Wisconsin being a hotspot for Lyme now (2025), back in the early 2000s the confirmed cases were a tenth of what they are today. In 2008, the surveillance case definition changed to include "probable cases," no doubt putting Lyme on the radar of more Wisconsin doctors, but this was still 2006-2007.

Bear with me as I get a little bit scientific. I promise it'll be brief and as concise as possible. I learned that doctors are expected to follow the CDC (Center for Disease Control) guidelines of a two-step testing process for Lyme, and both steps are required to confirm a positive result. An ELISA test is done first and, if negative, no further testing is recommended. If the ELISA is positive, a second test, the Western Blot, is done. These tests look for antibodies to the Lyme bacteria (the immune response to Lyme bacteria, Borrelia burgdorferi). The problem is that the sensitivity of the tests varies based on the stage of Lyme disease and where the bacteria are in the body. For example, they may only pick up 75 percent of neurologic Lyme disease. I also learned that some patients with chronic Lyme disease are found negative by ELISA yet test positive by Western Blot. For this reason, a prominent authority on Lyme now suggests that clinicians order both the ELISA and the Western blot at the same time.

Four weeks later, with continued joint pain, migraines, exhaustion, and brain fog, as well as new symptom of tinnitus (ringing in the ears), I trudged back to the neurologist for my recheck, food log in hand. Again, I brought up my desire to test for Lyme. *What's the harm in that, right?* My request was met with a clenched jaw from the doctor, but he agreed to order an ELISA test. The test came back negative. Knowing the shortcomings of the tests, I asked if he'd order a Western Blot test. He was quick to shut down that option with, "That's only done to confirm a positive ELISA." Again, feeling defeated, I filled the doctor's prescription for yet another medication that I hoped would prevent my migraines.

While following the science, I also discovered an alarming amount of controversy regarding testing, diagnosis, treatment, and long-term outcomes of Lyme, but this story is not about the details of these debates. I'm just here to tell my story. My reality.

I also found countless stories told by folks following a similar path as mine ... Feeling ill; lots of symptoms in multiple body systems; seeing a long list of doctors over multiple years; before finally being diagnosed with Lyme (and, often, other coinfections contracted from the same tick that caused Lyme) and embarking on their healing journey. The underlying common denominator was that these individuals persisted until they found the cause for their illness. I have no doubt some folks just give up and learn to live with their symptoms, perhaps even convinced by the medical community that it is "just in their head." I know this because there were times that I was tempted to just give up.

Along this medical journey, I began to realize that nobody knew my body better than me, and I needed to advocate for myself and make sure I was really being heard. I knew it was time to find a new neurologist. This time I chose a highly regarded medical center known for cutting-edge research. My new neurologist was beyond my expectations with her empathetic demeanor and her genuine interest in helping me get back to feeling good again. She truly heard me, and we put a plan in place to not only get to the bottom of my migraines but, hopefully, my other symptoms as well. By this time it was 2008, and I'd been experiencing heart palpitations on and off for months as well as episodes of a racing heart rate, lasting up to 30 minutes at times. This was disconcerting to say the least.

Via cardiac work up, we learned that I have a patent foramen ovale (PFO), a small hole in the heart that is supposed to close shortly after birth. One in four people have this condition and most never have clinical signs, but it can be an important finding for migraine sufferers because tiny clots can pass through this hole and go to the brain, triggering migraine symptoms. These are typically folks who get migraines with aura, and some, but not all, of my migraines occurred with aura. Much more research is needed in this area, but this was something to keep an eye on since the closure device was not yet FDA approved for use to prevent migraines. She started me on a medication to address my irregular heart rhythm as well as hopefully decrease my migraines, and it did help. Autonomic testing also revealed dysautonomia, a possible contributor to my dizziness. She added in a medication to address this, and I was finally starting to feel like my old (healthier) self again with less dizziness and fewer migraines. They weren't gone but they were better, and I was thrilled.

Over the next four years, symptoms continued to come and go, wax and wane, and new symptoms would be added into the mix as well. Tinnitus and ear pain flared up, and off I went to an otolaryngologist (ear, nose, and throat doctor), to find I had inflamed eustachian tubes and embarked on treatment for that. A string of odd, connective tissue, maladies starting to pop up including tears in my extensor carpi ulnaris (ECU) tendons of both wrists, tears in the triangular fribrocartilage complex (TFCC) of both wrists, and a shifting forward of my left collar bone, all of which caused significant pain and, in the case of my wrists, also caused neurological symptoms of prickling and weakness in my hands. In response to the ECU tear of my left wrist, I experienced complex regional pain syndrome, something I wouldn't wish on my worst enemy. Luckily, it was fleetingly brief. I just dealt with each issue as it arose including surgery to repair my right ECU tendon and TFCC. At the time, I wondered if Lyme might be at the root of my connective tissue injuries, but I purposely did not bring up this concern with my orthopedic surgeon. To be honest, I just didn't want my theory of this being "Lyme-related" to once again be scoffed at or dismissed by a doctor. He was an excellent surgeon, and I just wanted to get the repair done and move on with my life. I've since learned that the Lyme bacteria can invade and damage the connective tissue within tendons and ligaments, causing an inflammatory response that weakens their structure. This is particularly in cases where the infection is not treated promptly from what I understand. Perhaps my hunch was correct all along.

I don't want to give the impression that I was a suffering mess every day as that certainly wasn't the case. I went about my normal life and enjoyed an active lifestyle with my family. I just had to moderate my activity from time to time. I had the mindset that I was going to enjoy my life, and I focused on all the things I was thankful for every single day. This truly did help me navigate the darker days. I still yearned, though, for an answer as to why my body was having all these issues, and I wasn't even 50 yet.

The migrating joint pain and swelling and brain fog were still a big problem, along with significant fatigue, so my neurologist ordered another ELISA as well as a test for Epstein-Barr virus, both of which were negative. She then suggested I seek out an infectious disease doctor, as she felt that it was still possible that Lyme and/or coinfections might be at play. I put out feelers and found that many Lyme patients had great success with what were called "Lyme-literate" doctors. These were licensed physicians with the experience and knowledge as well as an open mind to help patients with symptoms as well as complications of Lyme disease, including coinfections. I found a Lyme-literate doctor in our network, a full day's drive there and back, and I had high hopes he could help me progress in the right direction. The doctor in me really wanted that diagnosis, that "positive test result," that validation, but I was prepared to consider diving into treatment for Lyme just based on my symptoms if he felt that was the way to go. One thing I learned, dealing with almost two decades of on and off symptoms, is that one can get quite frustrated, even desperate, and willing to try just about anything to feel normal again.

This doctor practiced complementary medicine along with traditional Western medicine, and this sounded like a holistic and smart direction for me. Along with medications, he used diet and dietary supplements and lifestyle to address his patients' issues.

He listened to my story, looked over my medical records, did a thorough exam, and then ordered a Western blot Lyme test which came back positive. Neither he, nor I, were surprised. I was both happy and sad at this news. Happy, to finally have a name for what was causing all the inflammation and issues in my body. Sad, because it had gone on for so long that more and more body systems were affected.

Borrelia burgdorferi, the spiral-shaped Lyme bacteria is stealthy and can hide out in many areas of the body. Researchers at various medical schools and clinics are starting to understand how this disease spreads so widely throughout the body, affecting any organ or system in the body. They're also starting to understand how it evades detection by the immune system and how it can hide out beyond the reach of drugs in some cases. A deep dive into all of this is beyond the scope of this chapter, but I've listed some resources at the end of the chapter if you're interested in finding out more about Lyme disease and the bacteria that causes it.

I embarked on my journey of healing that included multiple courses of antibiotics, herbal treatments, various supplements, and diet changes. I traveled back and forth across the state about once a month for two years and finally felt like myself again.

Medical conditions don't always fit into a neat little box. Lyme often has no bullseye rash. An ELISA may come back negative in a Lyme-positive person. One person's clinical signs may differ widely from another's. You know your body better than anyone else. If you feel you're not being heard, get a second opinion. Get a third opinion. Get as many professional opinions as you need until you find that doctor who really listens, who really empathizes, and who really wants to get to the bottom of the issue and help you heal. I was one of the lucky ones. Despite my long journey, I found two doctors who changed my world and my reality, and to whom I'm forever thankful.

There's a quote I learned in veterinary school that says, "When you hear hoofbeats, think of horses, not zebras." That might be true a lot of the time, but you know what? Sometimes it actually IS a zebra!

About Manette

Manette Kohler is a writer and a veterinarian in Southeast Wisconsin where she provides animal behavior consultations for cats and dogs. She also provides consultations to veterinarians regarding anxiety management for their patients. She's contributed as a freelance writer and columnist for nationally recognized *Dog World* magazine as well as Wisconsin-based *FETCH* magazine, and she also writes articles for *PetMD.com*. She speaks locally and regionally to veterinary audiences, pet owners, and other pet professionals, on animal behavior topics. In August 2018, she published her first children's book, *Bella's First Checkup*, a chapter book for young readers that teaches families how to raise a behaviorally healthy puppy and gives young readers an inside glimpse into a veterinary clinic. In November 2019, she was one of 20 authors who contributed personal stories to *Manifesting Your Dreams: Inspiring Words of Encouragement, Strength, and Perseverance*, a book of manifested dreams revealing how each author found their life's purpose through actual manifesting tools, the Law of Attraction, hard work, believing in themselves, or on the other side of trauma or tragedy. In this book she described how she manifested a whole new direction for her veterinary career, a thriving behavior veterinary practice. *Manifesting Your Dreams* reached #1 on Amazon, and it was featured twice in

O, The Oprah Magazine.

Dr. Kohler firmly believes that as we learn and grow intellectually and spiritually throughout our lives and careers, we continue to work toward our life's purpose. In this quest for fulfilling her life's purpose, she felt led to enroll in the Family Paws Parent Education program so that she could increase her knowledge regarding dog-and-baby/toddler dynamics within her clients' homes. As a licensed Family Paws Parent Educator, she can help families understand the many changing dynamics in a home that contains both dogs and babies or children. In this next phase of her career, she desires to take on more of a "teaching" role including providing "Dog-Baby" classes to families and consulting with veterinary clinics on important behavior topics including "Dog-Baby" safety and "Behavior Wellness" and how to incorporate this information into each clinic's philosophy. All her efforts work toward a single goal of making life better for people and their pets.

References and Resources:
1. *The Spectrum of Gastrointestinal Manifestations in Lyme Disease Fried, Martin; Abel, M; Pietruccha, D; Bal, A Journal of Pediatric Gastroenterology & Nutrition: Oct 1999-Vol 29-Issue 4 https://journals.lww.com/jpgn/fulltext/1999/10000/the_spectrum_of_gastrointestinal_manifestations_in.50.aspx*
2. *New York-Presbyterian/Columbia University Irving Medical Center https://www.columbia-lyme.org/lyme-disease*
3. *https://igenex.com*
4. *https://www.globallymealliance.org*
5. *https://medicine.yale.edu/intmed/raci/rheumatology/research/lyme/*
6. *https://www.scientificamerican.com/article/something-to-grapple-withhowwily-lyme-disease-prowls-the-body/*

ALISA STEVENS

*"Dreams do not have
an expiration date."*

– Cheryl Seagraves

Allow me to Reintroduce Myself

Music has always played an important part in my life. It has been my escape, a friend when I was feeling alone, and a connection to some amazing people. I am the person, when you are playing a musical trivia game, you want on your team!

A song came to mind, immediately, when I was asked to contribute a chapter to this book. I was at a Jay Z concert years ago, and I remember when his first song called "Public Service Announcement" began to play. It starts with these lyrics: "Allow me to reintroduce myself." Then he begins to talk about his life and accomplishments to date.

Introductions can be as simple as telling someone your name or more details including occupation, status (single, married, divorced), or where you live. As life changes, so can a person's introduction of themselves. It's like seeing someone you haven't connected with in years. Sometimes that person can be yourself. That person was me.

Growing up, I was always the kid who wanted peace and harmony. I wanted everyone to be happy, but most of all I wanted people to like me. I struggled with making friends because I was subjected to quite a bit of bullying. I had eye surgery to correct a lazy eye when I was five years old, and I had to wear special glass. Talk about being the subject of many jokes. I was called ugly, black (because of my dark complexion), and the best one I heard was "prison eyes" because the lenses in my glasses had lines that looked like bars. Let's just say that my early school years were not the best. My parents always said not to pay attention to those kids and that I was beautiful. I listened, but it was hard to believe. My older sister was the prettier one, and I was constantly reminded of that. I remember being introduced to a friend of my sister's and the person saying "Are you sure that's your whole sister? You look nothing like her." It was her tone that was so hurtful. Talk about a first impression. I felt like just running away, but I just smiled. I wish I would have stood up for myself, but of course I didn't want to rock the boat.

I figured that I would never be "pretty," but I could be smart so I just dove into my academics. I spent quite a bit of time in my room doing homework, reading, and of course listening to music. I could always find a song that said exactly what I was feeling.

At age 14, my parents decided to send me to modeling school. I still saw myself as the person my classmates described. I learned how to walk, pose, and apply makeup. Months passed and I graduated. My outside appearance started to change, but my inner negative thoughts remained the same. When I was 15, the modeling school did a pageant. Miss Teen Ohio was my first and my second pageant. I lost both times (one was to a very famous person). However, there was something about being on stage that made me feel like a different person. It was great, and I loved every minute of it!

I was even asked to go to New York City to give modeling a try, but my parents were not having that! They said, "Your education is more important, and who knows, you may not make it." I know that was their way of trying to protect their 16-year-old daughter, but hearing those words that I wouldn't make it sent me back to grade school and feeling never good enough.

I went back to focusing on my work and even started dating. Some dates were great while others just were not good for me—not good for my mental health. Because I was the person who wanted my peers to like me, I tolerated behavior that, when I look back now, was abusive. I gave up my voice to make my boyfriends happy.

I remember asking myself, *Who am I?*

I was so happy when I started college. It was a new start, and I thought it would be the place where I could find myself. I did find the "sister" I always wanted. It was the first person who appreciated me for me. We are still friends 35 years later.

After college, I started my first job as a nurse. Ironically, it was at the same hospital that my dad worked as a nurse for 29 years. He passed away from cancer six months before I graduated so working there was bittersweet. However, I was never happy working at the hospital, and I was restless and ready for a change.

I visited Chicago a year prior, and there was something about the city that I loved. The energy was infectious. My father used to say: "Sometimes you must lose yourself to find yourself." I didn't go to New York when I was younger, but what a better place to find myself than in Chicago? I remember telling my mom that I was quitting my job and moving. That went over well ... she was furious!! My parents were from the era of "stability." You get a job, get married, have a family, and you retire at that same job. I remember my mom saying that she couldn't understand why I would quit a great job with excellent benefits to move to huge crime-ridden city like Chicago. I tried to explain to her I saw better opportunities and a chance to be on my own. Saying that made the situation worse. I was always "the different one" because I always danced to the beat of my own drum (back to music, Lol). There were people back home who understood why and others who just didn't get it. I packed up my 1996 Saturn with everything I owned, which was primarily clothes, and headed west. My mom said to me before I pulled out of the driveway "Your key will always work." It was her way of wishing me the best but also doubting me. Off I went to find myself—to find my introduction.

The first few years I struggled. My first apartment was an old house converted into four separate units. It was okay but not a penthouse on Lake Shore Drive. I remember waking up to something crawling on my leg. I never lived with roaches growing up, and I certainly wasn't going to start at age 26! I immediately called my mom crying about my situation and she said, "You chose to move there." As harsh as it sounded, she was correct. I made the choice, and now I was going to figure it out. One call to the landlord and a couple of exterminations later, problem solved. My mom came to visit two years later. She didn't like my apartment, didn't like the area, and still questioned why I would leave Cleveland to move to Chicago.

After her visit, I had doubts. I started working two jobs seven days a week just to pay for my essentials. I was tired and frustrated. I also began to feel depressed. The place I came to find myself was the place that made me feel defeated. After three years, I made the decision to go back home. I started grad school and working part time. I also moved back in with my mom. I was dating someone in Chicago, so I was back at least twice a month. My mom was glad that I was back, and I was glad to see her, but something had changed. I changed. Cleveland was where I was born and raised but Chicago was me, it was home. So, I packed my car (this time it was a Nissan) and headed back to my now home. My mom said that it was time for me to settle myself down, but she also began to understand why I was going back.

I went back, but after a year things changed. I started a new job and that ended after six weeks. My relationship was rocky, and things seemed to go from bad to worse. I was able to find another job. My relationship, on the other hand, didn't survive.

Why did I go back?

I was no longer working two jobs, and that part was great, but I wasn't happy. Once again I was ready to make a change, but I was afraid because I doubted my choices. But something happened at this job that would change my life forever. I met a pharmaceutical rep who told me that I should investigate this option as a career. Change careers? Remember, stay at a job because you have great benefits and stability. I was so scared. I started dating someone new around the same time, and he knew that I was unhappy. I was seriously in turmoil. Then one night when I was driving home from work, one of my dad's favorite songs came on the radio, "Time After Time," by Cyndi Lauper.

"If your lost, you can look, and you will find me time after time. If

you fall, I will catch you, I'll be waiting ... time after time."

I pulled over and began to sob. It was like my dad was sending me an answer. Once again, music was my savior. I then decided it was time. Time for a completely new change.

I had several interviews, and I finally landed a job as a pharmaceutical rep. I was so thankful that my boyfriend, Kevin (now my husband), was by my side. Things began to change. I got married a couple of years later in 2004. We were living comfortably. Life was going okay, and then my life changed—again.

I was diagnosed with Lupus. I was always tired and began to have problems with swollen joints. There were times that walking became a challenge. I was glad to find an answer, but it wasn't the one I wanted. I was prescribed a cocktail of medications that didn't make me feel like myself, but I managed to put on a happy face. I didn't want anyone to know what was going on with me. When I told people about my diagnosis, I always heard, "You don't look sick." That would frustrate me even more. I just kept up my happy face.

When I became pregnant, I was overjoyed. But Lupus tried to take my happiness away from me. It made the pregnancy difficult. I was in and out of the hospital for months until I was put on bed rest for the last three months. My son was born at 37 weeks, but he was healthy and happy. We were happy. I began to have a Lupus flair a couple of days after he was born. It was the worst; however, it wasn't as bad as finding out that my husband was going to be deployed to Iraq in 30 days.

My mom packed up her house and moved in with me and her grandson while Kevin was deployed. There was no way that I could work full time (out of town travel is part of the job) and take care of my baby. I was a new mom with anxiety off the charts because my husband was off to war. My mom was a saint, and I couldn't have done it without her. Our relationship strengthened during this time. Kevin came back nine months later. I was so glad that he was safe and back home. During that period of deployment, I changed. This was the darkest period of my life.

To others, we were doing fine. Great house, beautiful son, a couple of amazing animals (a dog and two cats), but inside I was a mess. I was still taking a cocktail of medications, and my depression and anxiety were worse. I started to feel angry at the world. I was mad at the United States Navy for taking my husband away from me when I needed him the most. I was mad at Lupus for making me feel horrible—so badly that I didn't have the energy to take care of my son and be a great wife. I didn't like the person I had become, and I didn't think that anyone liked me either. I didn't think my family deserved to live with an angry person. In my head, I was ready to end it all.

I was admitted to the hospital, and those were the longest three days of my life. I couldn't have shoelaces or clothing with ties. I was in a room with a bed and blanket. I had to brush my teeth and take a shower with someone watching. I remember doing a clinical rotation on a psychiatric unit while in school. I never thought that I would some day be a patient. I told the therapist I was angry at my husband for taking me to the hospital. The therapist said, "You should thank him because he loves you and saved your life." That sentence started the change in me. I started to look at myself and get to the root of my anger.

I was able to receive calls while in the hospital, and the common theme I was told was this: "There is nothing wrong with you; it is your husband who put you there. You don't belong there." I was at a place where I couldn't, and I didn't feel like fighting. But those conversations made me begin to figure out why I was so angry. I was angry at myself for not speaking up for myself. Angry for letting others put me down. Angry for not letting myself be me. That was a turning point. I was ready to let go and find me. I needed a reintroduction.

I found a therapist who helped me to understand my anger. I remember leaving sessions feeling worse than when I arrived, but I realized that was part of the journey. I also found out I was pregnant with my second child during this time. It was challenging, but when I gave birth to my daughter, we were elated. We had our son and daughter. This made me more determined to be a perfect mom. That raised my anxiety. I am so blessed that I had an amazing support system to help me, but I also had to help myself.

I finally came to a place in my life where I was moving forward. There was only one thing that I had always wanted to revisit and that was doing another pageant. Soon I met a woman who completed her first pageant at the age of 53. Her energy and spirit lit a fire in me.

I had the dream, but I just didn't know if I could do it. Then I started to think, *What is truly holding me back from pursuing my dream?* It was me! Yes me!

During the same time, my mom was diagnosed with cancer. She endured surgery and then lived with us. I was her caregiver along with my husband.

Could I really take on this dream? And the answer was yes. I was appointed Mrs. Illinois American 2021! It was such an amazing feeling when I came home with the crown, and I was able to share it with my mom. That was a special moment.

Next was the national pageant. Mom was able to go gown shopping with me, and she helped me pick out my wardrobe for this big pageant. I didn't win the national title but won something else. I was proud of myself for following my dreams. I loved talking with young girls and letting them know they are all queens. Giving back was the best feeling.

My mom passed away in May of 2022. I was so thankful that she was able to live with us, and I was able to be a part of her care. The last conversation was a little over a week before her passing. I was able to thank her for everything she has given me. I let her know that she was a great mom, and I am truly thankful for her. She took my hand and said that she was so proud of the woman I had become and thanked me for everything I did for her. That was incredibly healing for me. I know that she is reunited with my dad in heaven.

Through this journey, I found my voice, my faith, and most importantly, I am finding me.

So, allow me to reintroduce myself. I am a caring, loving, sensitive, music-loving, animal-loving, work-in-progress woman, wife, and mom. And I am loving every part of the new construction.

About Alisa

Alisa Stevens is an award-winning Pharmaceutical Account Manager for a Biotech company. She was Mrs. Illinois American 2021 and United States of America Mrs. Illinois 2023. She placed in the Top 12 at nationals and was the oldest contestant to make the top at age 53. Her platform is called "The Lupus Thriver," and she has had the opportunity to share her journey with Lupus to others through volunteering with The Lupus Society of Illinois, The Lupus Foundation of America, and speaking nationally and internationally. She is married to Kevin (21 years) and has two children (Christian-19, Victoria-13).

She enjoys traveling, exercising, and being a self-proclaimed foodie.

One of Alisa's final photos she took as United States of America Mrs. Illinois.

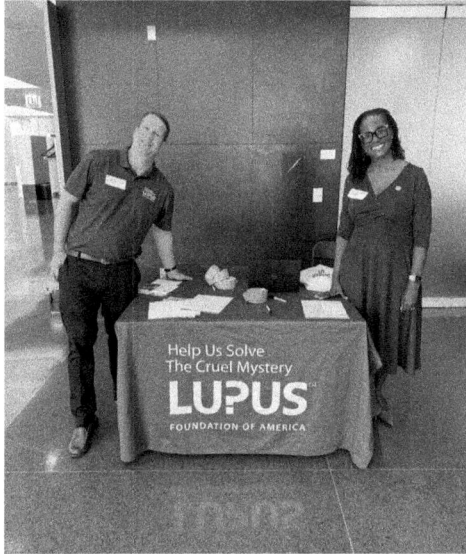

*Working at a conference as an ambassador
for the Lupus Foundation of America.*

Alisa's parents on their wedding day.

ADAM A. MEYERS

"God has given each of you a gift from his great variety of spiritual gifts. Use them well to serve one another."

– 1 John 4:10

If I'd Only Broken My Leg

On Friday, April 8, 2016, at 5:15 p.m., I was involved in a critical incident when I used deadly force against someone who armed themselves with a hatchet inside a busy department store. This was the first officer involved shooting in the history of the police department. This person died. When I was told this person died I thought to myself, *This changes everything. I haven't just shot someone. I had taken someone's life. The life of a mother, a daughter, a cousin, someone's friend. My critical incident was different now. I had no idea what was in store for me. I felt many different emotions but mainly numb. And the numbness would not go away for years.*

I faced many personal and professional mental health challenges. It has taken me several years to get back on track. There are many different coping strategies people may use after experiencing trauma. They may be good and healthy, or they may be bad and unhealthy. My coping strategies were bad, unhealthy, self-destructive, and dangerous. I used alcohol, marijuana, casual sex, and self-harm as some of my poor coping strategies for years.

A year after my critical incident, I was promoted to the police department's first detective. I investigated an attempted homicide, a quadruple homicide where the suspect died by suicide, suicides, vehicular homicide where three girl scouts and their troop leader were killed after being ran over by a vehicle operated by a driver who was huffing. I chased a bank robber for 18 miles with speeds exceeding 100 mph; they previously robbed three other banks. I investigated a shooting that occurred in the parking lot of the same retail store where my critical incident occurred. I experienced cumulative trauma for years after that.

My poor coping strategies easily put my relationships, job, and health at risk, but I did not care. I wanted to escape from what I was feeling. I wanted to numb my emotions, my thoughts, my body, and any memory of taking someone's life. I wanted to feel better even if only momentarily. I was selfish, reckless, and I did not care how my self-destructive and dangerous behavior may have affected my family, friends, children, co-workers, and the public.

One of my many poor coping strategies I used was abusing alcohol. Prior to the incident, I collected wine and enjoyed a glass every now and then. However, afterward, I began abusing liquor, mainly whiskey and the cheapest vodka I could get my hands on. I would consume whiskey and vodka straight from their bottles, on the rocks, or I would create my own cocktails by combining over-the-counter liquid sleeping or liquid allergy medicines. There were times I would mix in whatever leftover prescription medicines I had in the medicine cabinet, and it didn't matter if they were prescribed to me or someone else.

For example, I remember one instance when my oldest daughter, Sydney, had leftover prescribed liquid cough medicine containing codeine. I combined whatever was remaining in the bottle with a glass of wine. I was on a prescribed anti-depressant called Venlafaxine while I was abusing alcohol. The label on this medication specifically stated, "DO NOT DRINK ALCOHOLIC BEVERAGES WHILE TAKING THIS MEDICATION." A warning label did not deter me from abusing alcohol. I very well could have blacked out and never woken up from consuming these dangerous cocktails, but at the time

I didn't care. Abusing alcohol may have been a quick fix, but it caused me even more stress, anxiety, and depression.

In September 2018, I drafted a written contract with myself to not consume alcohol. I don't remember writing the contract, but there was something inside of me that recognized how self-destructive alcohol was. This contract quickly became null and void because it took me another three years to reduce my alcohol consumption.

There were many times I did not want to go to work. This was not because I had other plans or that I was hungover from consuming too much alcohol but because I just wanted to stay at home and isolate myself from the world. I wanted to lock all the doors of my home, close all the curtains, and shut everyone out of my life—which I did many times for many years. Sure, I called in sick from time to time, but on one occasion, I intentionally injured myself so that I didn't have to work. I used an old 12-inch adjustable steel wrench to cause superficial injuries to my left knee. I struck my knee a dozen or more times, enough to cause redness, abrasions, and bruising, and limped into the local emergency room. I explained to the doctor and nurses that I had tripped and fallen down walking out of the back door of my house and struck my knee on the steel covering of an underground septic tank. My story was believable enough. I received X-rays of my knee, a prescription for pain medication, and crutches. I was discharged from the emergency room with a doctor's letter releasing me from work for about one week. This occurred during a busy holiday work week. Although I was not able to truly celebrate the holiday, this deception got me out of work, and I was able to enjoy the time alone at home.

Another way I was able to get out of working was to intentionally make myself sick. I remember while taking a shower before my shift I was thinking about the many ways I could get out of work. While I was putting on my ballistic vest, uniform, boots, and duty belt, I thought to myself, *I'm going to get out of working by binge eating and making myself vomit.* I constructed a plan to visit the local Burger King drive-thru while traveling to work. I ordered a bunch of breakfast food and made sure I washed it down with a large soda and a large orange juice. I needed to make sure I added beverages to my breakfast buffet to ensure it would all come back up easier.

I continued to travel to work and passed the local McDonald's and thought, *Two is better than one.* I quickly binged what I purchased and proceeded to work.

I arrived at work, and upon exiting my vehicle I masterfully played the role of the "sick employee." I walked into the police department and made myself vomit in the bathroom. I made sure the bathroom door remained open so that anyone walking by could see or hear me. I made sure that not all my vomit made it into the toilet and landed on the floor for added effect. I was immediately sent home.

Another way I dangerously coped was by drinking and driving. Prior to attending any type of social event, even as simple as going to the grocery store, I would consume alcohol. I would travel to a nearby gas station and purchase many small bottles of liquor containing about 1.5 ounces of whiskey, vodka, or whatever I could afford at the time. I would immediately consume the alcohol in my vehicle prior to traveling to my destination. I tossed the empty bottles in the back of my vehicle or out the window while I was driving. I would rationalize that it would take about 30 minutes for me to feel the effects of the alcohol, and by the time I was impaired I would have arrived at my destination. I was very fortunate that I was not arrested for drinking and driving or even worse, killing someone.

Another example of risky and reckless behavior that I am still ashamed of today involved alcohol, operating while impaired, and my youngest daughter, Peyton. I was consuming alcohol late morning into the early afternoon and had to pick up Peyton at her mother's (my ex-wife's) house and take her to gymnastics. As we drove to gymnastics, I took the wrong turn and drove 21 miles out of the way. I drove for 61 miles impaired by alcohol. For 46 miles, I had my daughter with me.

I used casual sex as a coping strategy and to distract myself from my emotional discomfort and pain as well. I would meet women and sometimes within 30 minutes, we would have sex. This caused me more stress and anxiety than anything else. Sure, I felt great during sex, but it caused me more harm than good. I was constantly worried about pregnancy and contracting a sexually transmitted disease. Although this type of sexual behavior was risky, self-destructive, and caused me stress and anxiety, it was not enough to convince me to stop. I wanted an instant feel-good escape from my life and casual sex provided that for me. I remember lying in bed with a woman after having sex and thinking, *What am I doing?* I felt ashamed and empty; I didn't like myself very much.

Another dangerous and unexplainable way I coped was by putting my duty weapon to my head. During my critical incident, it was a Glock 22 Gen 4 – .40 Caliber. I put this weapon to my head at least a dozen times. Sometimes I even placed the barrel in my mouth. I would always remove the magazine, but for those of you who are not familiar with a Glock, if you don't rack the slide and remove the round from the chamber (barrel) it will still discharge a round. I very easily could have accidentally killed myself. My rationalization was that I simply wanted to hear and feel the metallic click of the trigger being pulled while the barrel of the gun was resting against my right temple. I did this while I was under the influence of alcohol. I still do not truly understand why I did this, and sometimes wonder how many times it happened while I was excessively consuming alcohol. I am very fortunate to be alive.

I always lived in a state of hypervigilance. I always thought something bad was going to happen to me. I transitioned from sleeping in my bedroom to sleeping on the couch in my living room. My reasoning was that I wouldn't be able to see what was occurring within my house if I was isolated to my bedroom. I needed to be ready to react quickly if someone forced entry into my home and wanted to harm my children or me. I always had my duty or off-duty weapon within reach while sleeping on the couch, cutting the grass, washing the dishes, or simply relaxing. If you can even call that relaxing.

I suffered in silence for many years after my critical incident and I am ashamed for the ways I poorly coped. I still feel shame today for the ways I coped and treated people. I find it hard to believe that nobody realized or even had a gut feeling that I was not doing well. *I could not have been that good at hiding my poor coping strategies, or was I?* I have always wondered if people were slowly watching me self-destruct because they did not know what to say to me, how to help me, or they simply did not want to get involved.

In March 2020, I was hired by a sheriff's department in southeastern Wisconsin in a county where most of my family resided. My plan was to continue working in the law enforcement profession. Then, after about 10 years, I would retire near my family. I was hired by the sheriff's department through a lateral transfer and received a significant pay increase. I immediately purchased a house before beginning my new job. I thought I was mentally and physically healthy, but I was wrong. I resigned 40 days after being sworn in. I hadn't even paid my first mortgage payment.

I experienced a lot of the same types of mental health struggles on and off duty while at the sheriff's department because I never addressed past struggles while employed at my previous police department. Those struggles just don't simply go away; they were with me wherever I went. I could have moved clear across the world, and I would have experienced the same struggles. I was constantly being triggered at work: traffic crashes, suicides, drunk drivers, the beginning of COVID-19, and trying to prove to my field training officer, lieutenant, and captain that I could do the job. I thought I was losing my mind. I was having difficulties doing simple tasks like report writing, operating a patrol vehicle, navigation, speaking with people in person or on the telephone, and talking on the police radio.

My resignation explained, "Due to my increased PTSD from my 2016 officer involved shooting, it is not safe for me to continues as a law enforcement officer."

In June 2020, I was hired back as a detective at the police department where my shooting occurred. I interviewed with the police committee and completed basic pre-employment screenings, but I didn't have a psychological exam. I was surprised because my resignation letter from the sheriff's department specifically stated I was resigning because of my increased PTSD from my shooting and that I didn't believe it was safe for me to be a law enforcement officer. But I was very grateful to be hired back where I truly enjoyed working.

In December 2021, I experienced a panic attack while attending an active-shooter training at a middle school. A few days later, on New Years Eve, before beginning my shift, I explained to my sergeant and later to my chief, that I wanted to quit. I was relieved from my shift and sent home. Looking back, I don't think I wanted to quit being a police officer, I think I wanted to quit the poor coping strategies that plagued me for years. Coping poorly didn't help me, and I was ready to get real help. I wanted to get better.

In January 2022, I had a psychological examination. I needed to be evaluated and deemed fit for duty before returning to work. The assessment lasted seven hours and resulted in a nine-page mental health diagnoses. I was thinking about not being honest during the evaluation, and I thought I could beat the evaluation. I chose to be honest with the evaluation and myself; that was the only way I would receive the type of help I need.

I was diagnosed with Major Depressive Disorder, PTSD, Trauma and Stressor Related Disorder with Acute Stress, Panic, and Dissociative features in January 2022 by the police department's psychologist and deemed Unfit for Duty. This meant I was not able to return to work and needed to undergo intense Psychotherapy, Eye Movement Desensitization and Reprocessing (EMDR), Biofeedback, and Dialectical Behavior Therapy (DBT). This occurred two to three times a week.

The police department's psychologist recommended six months short-term disability with a goal of returning for duty in August 2022. This did not occur, as I was terminated after the police department wouldn't extend my 90-day leave of absence. I feel the police department gave up on me. They abandoned me. They washed their hands of me, and I wasn't their problem anymore.

I experienced financial hardship for almost two years after my 90-day leave of absence. My credit score plunged into the 500s. My self-esteem, confidence, and depression also plunged. I was never able to recover financially and filed for bankruptcy in 2024.

In April 2022, during my leave of absence, I was placed on a safety plan because of my suicidal ideations. The psychological exam discovered an elevated risk of harm to myself. I agreed to not engage in self-harm, destructive or life-threatening behavior, or any other high-risk behavior to myself or others. I don't remember ever wanting to die by suicide, but my behaviors were very evident that

I was heading in that direction. There were many nights I would pray to God that he would not let me wake up in the morning. Fortunately, I had to go through every second, minute, hour, day, and year in order to get better. This made me stronger, and I was able to better understand my mental health. I truly believe that God put me through all of these struggles so I could help others.

Also, in April 2022, I received an official order from the police department instructing me to immediately cease and desist disseminating personal testimonies regarding my mental health as it related to my critical incident. I resided outside of the jurisdiction in which I was a police officer. A police supervisor traveled to my residence in a fully marked patrol vehicle, wearing their patrol uniform. They parked in my driveway and issued me the order. I felt intimidated and confused as I had been speaking publicly about my critical incident and mental health experience since 2019. I was even encouraged to continue to speak, while on my leave of absence, in which I did at a local sheriff's department.

I was finally undergoing the right kind of mental health treatment that helped me heal and understand why I adapted to poor coping strategies to self-medicate myself. I only wished I would have begun this therapy five years earlier, but it was better later than never.

In May 2022, I met with my chief about the status of my leave of absence. The chief informed me that my leave of absence was not being extended. He literally had two letters for me to choose from regarding the end of my employment with the police department. One was a termination letter and the other was a resignation letter. I refused to resign and was terminated. I was escorted into my detective office, turned in my equipment, collected my belongings, and escorted out of the police department. The last thing that was said to me —by the same supervisor who issued me the official order to cease and desist speaking about my mental health—was that maybe I could speak more about my mental health since I wasn't employed by the police department.

What next? I had been diagnosed with a mental illness, on a safety plan, and terminated from a job I'd wanted since I was a little boy. I had been employed with the police department for 14 years.

What if I would have gone home and died by suicide? It was almost as if the police department washed their hands of me. I was ghosted by my co-workers. These were the same people who not only knew me professionally but personally. They knew my family, my daughters, helped me move, and even came to me with their own personal and professional mental health challenges. I would have died for them. I felt hopeless, helpless, and abandoned. I was devastated.

Within weeks of being terminated, I moved three and a half hours away. I had lived in the area for 16 years, but I needed a new beginning. I needed to distance myself from my trauma; if that's even possible. I moved into the basement of my sister Amy's house. I was not taking medication or seeing a counselor, and I wouldn't start again for several months after I moved. I knew that I needed to, but I would not take the initial step to get back into therapy. Thankfully, my sister was researching and communicating with therapists without my knowledge. Amy could see I needed help and wanted me to get better. I am truly blessed by her. I began psychotherapy and EMDR again and was prescribed Lexparo.

I never thought I was going to be a police officer again. But. I realized that I wanted to return to law enforcement after attending therapy and getting healthy. I have a lot of training and experience that I believed would benefit not only myself, but rookie and veteran officers. I also wanted to share my experience in hopes that it would help bridge that gap between law enforcement and those in the community struggling with mental health.

In 2023, I began applying to become a police officer about one year after being terminated. I attended many interviews and underwent background checks, pre-employment medical screenings, and psychological evaluations. In one week, I received three conditional offers of employment from three different law enforcement agencies in three different Wisconsin counties. These conditional offers of employment boosted my confidence and confirmed that I wanted to become a police officer again. The only way I could have ever accomplished this was because I put myself first and focused on improving my mental and physical health. I continued to receive additional offers of employment even after being hired.

In July 2024, I filed bankruptcy. I tried for almost two years after being on a leave of absence and terminated to catch up with my financial responsibilities, but I could not get back on track. I didn't receive workers' compensation while on leave of absence, and that crushed my credit and put me in a deep financial hole that I couldn't dig myself out of. *Did I not receive workers' compensation because I was on a leave of absence for mental health?* I have no doubt in my mind that if I would have broken my leg or physically injured myself in any other way I would have received worker's compensation. *Why didn't the police department assist me financially?* This was a work-related injury.

I had to apply for and eventually received unemployment, but it was not enough to cover all of my financial obligations. So, while I was undergoing intensive mental health treatment in hopes that I would return to work in August 2022, I had to apply for and interview with other jobs in order to maintain my unemployment benefits. What I fiasco! I don't understand why for years I paid for health insurance to help me during a health emergency, but when I needed help I was denied.

I continue weekly therapy that includes in-person psychotherapy and Eye Movement Desensitization and Reprocessing (EMDR). I am also prescribed Lexapro (20mg) and Propranolol (as needed) which are medications that help me with my depression and generalized anxiety.

In 2019, I began speaking publicly about my personal and professional mental health challenges related to my 2016 critical incident. I established Stop The Threat – Stop The Stigma, LLC in December 2020. My overall goal for establishing Stop The Threat – Stop The Stigma and speaking about my critical incident is to promote Law Enforcement Wellness in hopes that I will inspire other law enforcement officers and those in the public safety professional to speak about and seek help for their mental health.

I am currently a captain with the same police department that hired me in 2023. I am where I am today because I finally put myself first and chose to fight my way back to my passion of being a law enforcement officer. I chose to finally take back my power. I have been able to move past my poor coping strategies because of the support I received from my family, friends, therapists, former girlfriend, co-workers, and complete strangers. I would not be where I am today without them. I am very grateful and will never be able to put into words how much their support means to me.

Please reach out for help if you or someone you believe is struggling with their mental health. I know it may feel awkward or uncomfortable, but most people will not admit they are struggling, and most people will not reach out for help. You could be a light during a very dark time in their life.

Remember, it's okay to talk about your mental health. You are not alone. Please don't suffer in silence.

About Adam

Adam Meyers is a captain with the Hartford Township Police Department in Washington County, Wisconsin. Adam began his Law Enforcement career in 2001, after five years as an active-duty United States Army Military Policeman. Adam has been a police chief, detective, major crimes evidence technician/custodian, field training officer, and is a police instructor in Firearms, Professional Communication Skills, Scenario, and Officer Wellness. Prior to and during Adam's law enforcement career, he spent about 10 years working with Behavioral Health Services in Southeastern and Northwestern Wisconsin and for hospitals with Behavioral Health Units. Adam is a certified peer specialist and is also employed at the STRONG Milwaukee Center in Milwaukee, Wisconsin, which is a day treatment for children and adolescents who have significant mental health and behavioral health issues. As a Mental Health Advocate, Adam takes every opportunity to speak about his personal and professional challenges with mental health after his 2016 on-duty deadly shooting.

More about Stop The Threat – Stop The Stigma

Adam is the Founder of Stop The Threat – Stop The Stigma. Adam says his overall goal for establishing Stop The Threat – Stop The Stigma and speaking about his critical incident is to promote

Law Enforcement Wellness and inspire other Law Enforcement Professionals, and those who work in the law enforcement profession, to speak about their own mental health.

Stop The Threat – Stop The Stigma provides general information and online resources for Law Enforcement Professionals and other Public Safety Professionals who may be struggling with their mental health.

The issue of mental health in law enforcement is prominent everywhere, and it is an issue that commands a new perspective. Without advocacy and awareness, we will continue to wait until the crisis stage to address mental health for law enforcement officers. By then it may be too late.

The law enforcement culture tends to expect law enforcement officers to resist normal psychological responses to tragedies or critical incidents. This combination creates the perfect storm when law enforcement officers are not provided with the tools to deal with the effects of the profession. For more information, please visit: stopthethreatstopthestigma.org

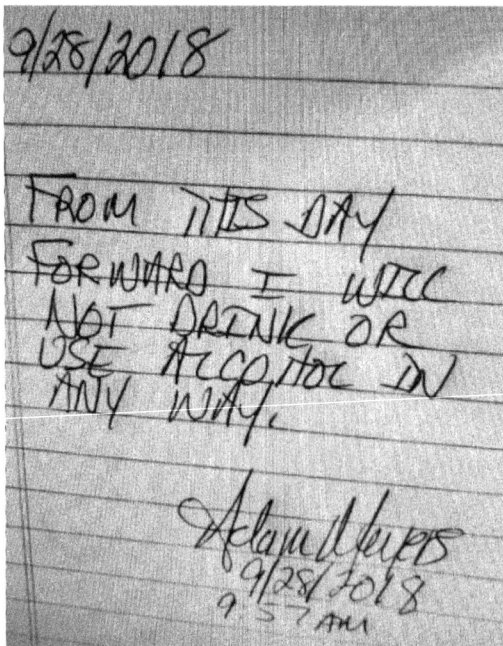

Scenes from Adam's home while he was depressed and poorly coping.

Hate mail.

Suicidal thoughts.

SHARON MANIACI

"You have always had the power my dear, you just had to learn it for yourself."
– Glinda from Wizard of Oz

From Victim to Survivor

*K*ris woke in a sweat, tears dried on her face. As she reached over to her left, feeling the cold empty spot, she remembered that her husband, Tony, was on deployment again. The dreams, nightmares really, had started again, this time more vividly and fluid than before. Flashes of memories had been resurfacing in her mind as if they were trailers for a movie that were playing over the last couple of years. This night was the worst. Kris had left the St. Louis Police Department 16 years ago and worked for a small town for a few months before she met Tony. Then she ended up moving to San Diego and getting married. Now, back in Virginia for the last 12 years, her job with the police department felt like a lifetime ago ... until the nightmares began.

This is how I began to take my power back. In 2021, I published my first solo book titled *The Ride Home*. A couple of years prior to that, I met with Marla McKenna before a Rick Springfield concert in New Buffalo, Michigan. I had asked her to sign one of her children's books for my goddaughter's baby girl. During our conversation, I had mentioned that I practiced reiki. Marla was intrigued and wanted to know more about it. It was then that I first began telling her my story.

I explained that in 2013, I started recovering memories of a brutal sexual assault, perpetrated by two of my co-workers when I was an officer with the St. Louis Metropolitan Police Department. During those long drives back and forth to St. Louis, filing charges, and talking to the police, my younger sister, Tricia, told me I should have reiki done at a shop in the area she would frequent. I attended several reiki sessions in the short time I was home which did a world of good for me. It helped me so much that I attended classes to become a certified reiki practitioner.

Marla and I continued our conversation, and I told her that another way for me to heal between reiki, therapy, and yoga (at my therapist's suggestion), was to write everything in a journal. The process of writing began at the onset of the nightmares. Just as I described in my book *The Ride Home* with the character Kris, I woke early in the morning from the nightmares and wrote down everything I could remember. I also told Marla about the process of going back and forth between St. Louis and Virginia, the exhaustion I was feeling with the trips, and reliving everything I remembered almost 20 years after the fact.

My first trip back to St. Louis, pertaining to the recovered memories, followed with a letter from a sergeant in the Sex Crimes Unit of the St. Louis Police Department. However, before I received the letter from him, I had sent a letter to three divisions in the police department: one was sent to the chief of police, one to the Internal Affairs Division and one to the Board of Police Commissioners. It outlined the events that occurred at the hands of two men I worked with—in as much detail as I could remember at that time. I sent the letters certified so I would know they had been delivered.

I wasn't sure what would happen when I sent those letters, at least until I received the letter from the Sex Crimes Unit. It too had been sent to me certified. I remember the day very clearly. I was on the patio in the backyard, grilling dinner. When the doorbell rang, my husband, David, was inside so he answered the door. I could hear him talking but couldn't pinpoint what was being said. He came to the patio door and told me I had to sign for something that came in the mail. I quickly signed for the letter, staring at the address. I sat down at the kitchen table and opened it.

"Dear Mrs. Melton, I am Sgt Pollard with the St. Louis Police Sex Crimes Unit. We are in receipt of your letter outlining the events that occurred in 1996, and we would like to hear from you regarding circumstances surrounding them." I was in complete disbelief, and my first reaction was to cry. I looked at David and said, "They believe me. They want to talk to me and didn't ignore me." This moment was huge for me. I finally felt like I was getting somewhere. Just to be able to tell someone within the department, someone who wanted to know, was extremely important. I called the next day and spoke with the sergeant. He told me I didn't have to make any decisions about filing charges right away but wanted to know if I wanted to talk over the phone or if I would be in town any time soon. I told him when I would be in town next, and I set up a time to meet with him.

I made a trip home that July for Mom and Dad's 70th birthday party. I was going to be there for about two weeks, so I had plenty of time before and after the interview for anything that may need to be done with the department. The party was first, and the following Monday I went down to the police headquarters. I was fortunate that my younger sister and best friend, Kate, went with me. I didn't actually meet with the sergeant, but a detective with the unit and a sexual assault advocate were there for me as well. The interview went fine, and I was asked if I wanted to set up a time to view photographs, which police officers refer to as a "six pack." At this point during the process, I still had not remembered the name of the other officer who raped me. I set the meeting time to view the line-up photographs for that coming Wednesday.

This time I had my mom with me, but like my sister and friend, she was not allowed to come into the interview room. The photographs I was provided to view were of police officers, but the pictures were taken in their uniforms, just before we graduated from the academy and the caps worn in the photos prevented me from seeing their eyes very well. In hindsight, this didn't really matter because of the six photos I was provided, three of them did not even come close to matching the description I had given them of that second officer. I may not have remembered his name, but I did remember what he looked like. None of the photos were of the second officer. It was a complete let down for me. I told the detective that if I could view the class pictures from the classes that graduated the academy in 1996, I would be able to pick him out of the picture, despite there being so many people in them.

The detective told me he would talk to the sergeant about accessing the class photos and get back to me. When I heard from the detective that Friday, I thought it was about viewing the class photos. I was wrong. The detective told me that the statute of limitations for filing criminal charges of rape passed in 1999, and that it was only a three-year statute at the time. This meant that I would have had to remember the incident by 1999. When the detective called, I was out with my mom and dad, and I had just parked the car. When the call came in, they stepped out of the car and when I hung up, I laid my head on the steering wheel and sobbed. I couldn't stop. I was so devastated, I couldn't think, I couldn't breathe, and I started to hyperventilate. My mom opened the door and held me. I felt like I was a small child again and had just fallen down. They offered to drive me back to the house, but I told them no. I needed to keep going with my plans for the day. There was nothing I could do to change the situation and sitting at the house wouldn't help me.

We went about our day, and I went about my plans for the remainder of my visit. On my way back to Virginia, I spoke with the sergeant from the sex crimes unit who told me they brought the officer whose name I knew in for questioning. They told him they wanted to know his side of the story, but the high-priced attorney he brought with him wouldn't allow him to say one word. The attorney wanted to know who filed the charges, but the sergeant would not tell him because the officer would not talk. The sergeant said nobody was talking. The officer and his attorney walked out of there without saying a word.

Now that the statute of limitations was over and I could not remember the name of the second officer and the other wasn't saying a word, my only option was to look into filing a civil suit. I was not giving up on finding the name of the other police officer though. I sent an email to a former boyfriend I worked with in the city asking him to call me from a secure phone line. I knew he worked for the federal government, and calls would be recorded. After a conversation with him, I was able to confirm the first name of the officer and had the first letter of the last name. He wasn't able to remember his entire last name, but he knew what it could have been, and I had more to go on at this point.

On my next trip to St. Louis, I met with an academy classmate. I had not seen her in almost 20 years, but we reconnected through my older sister and Facebook. We planned to meet when I got into town. My next call was to set up a meeting with an attorney to file a civil suit. I really wanted to have the name of the second officer, but I wasn't left with much choice. The academy classmate I would be meeting said she would go to the academy library with me to look at the class photos. These were the same academy class photos that I was told by the sex crimes sergeant were almost impossible to get a hold of. His words to me were: "It would be easier getting into Ft. Knox." This was completely untrue. The library was open to anyone, and the photos were part of the library. This was also one of my first indications that despite being told nobody knew my one attacker, but only knew of him and of course his father who was a captain of the police department. I figured out that this was a lie too. The "six pack" line-up of photos was so far off the mark, and it was obvious to me that day. I distinctly remember crying after viewing those six photos, but my tears were out of frustration of not being able to pick the perpetrator, not out of anger or hurt for the situation. Now knowing that it absolutely would have been possible to get those class photos frustrated me even more.

A couple of days after being in St. Louis, I picked up my friend, and we drove down to the St. Louis Police Academy building and went to the library. We were looking for the class photos of all the classes that graduated in 1996. There were six classes that year, so we would each take one of a set of two. On the back of each photo was a listing of each officer by row. I would only look at the photo and then let her look at the names. The first two sets were a wash. There wasn't even an officer with the first name of my second rapist. As we were looking at the last set of photos, she looked at hers and said "He isn't in here either," but as I looked at mine, I put my finger on the photo of the face of an officer and slowly slid it over to her. She replaced her finger with mine so she could keep the spot marked. Then she flipped the photo over and matched the face with the name.

"You did it. You found him."

Sixteen years later, I finally saw the face of the second man who raped me. I kept looking at the photo, at his face, and couldn't say anything. I felt my friend's hand on my shoulder, gently squeezing it. "I've got you Maniaci." The academy librarian came back to the window and asked if we found what we were looking for. I still wasn't able to speak, so my friend told her we had and thanked her so much for helping us. As we left the building and were walking down the stairs, I could feel my knees start to buckle under me, but as she said before, my friend had me. She was there to make sure I didn't fall.

I met with the attorney, who had asked me to pick up a copy of the police report I filed. When we started talking, I explained to him that I wasn't able to get a copy of the report because it had been flagged for some reason. Despite the Sunshine Law, which provides access to police reports, especially to the people who filed them, I was told by the clerk of the records department she couldn't give me a copy of the report. It just so happened that clerk was someone who worked in the station, at the time, with not only me but my two rapists. This was clue number two that there was something more going on behind the scenes. The attorney told me he would make some calls and get the report for me. I spoke with the attorney for about an hour, and he let me know what else he would need as far as paperwork and other items to get the suit started. He needed a copy of my journal entry from when I recovered the memories and documentation from my therapists.

I returned home to Virginia and began gathering the things I was requested to send to the attorney. After a week of being home, I heard from the attorney, who informed me that he wasn't going to be able to represent me in a civil suit because he didn't think I had a case. He told me he would send me a copy of the police report, but there was nothing further he could do. At the same time all of this was happening, the therapist I had seen in Richmond told me she no longer had any of my therapy notes and that the work she and I did, didn't involve any of my memory recovery, so she shredded everything. The hits just kept coming, and it was starting to feel more and more like a cover-up. The attorney I had seen was originally an attorney in the Circuit Attorney's Office for the City of St. Louis, and it hadn't worried me … until now. I couldn't figure out what was happening with the therapist though, and that was concerning.

I was back to seeing Tina, the therapist I originally worked with in 2009, and she was able to get the records from the other therapist. Even she did not understand why the notes and records would have been destroyed because they are supposed to be kept for a certain period of time following the discharge of their client. While she was working on that, I was working on finding another attorney. I found one through the friend of my older sister, so I contacted the attorney through email. I heard back from her fairly quickly, and I called her later that week. I made an appointment to sit down with them and see what they had to say. This meant I would make yet another trip to St. Louis. I did not have much time though because the window for filing a civil suit was approaching within the upcoming months.

When I made it to St. Louis, I had a copy of the police report, all of my journal entries, and a copy of the notes that had not actually been shredded by the therapist in Richmond. Tina said she would provide anything they needed from her. My older sister went with me to meet the attorneys. Like all the other meetings I'd had, she was not allowed in the room but having the support waiting for me was important. I met with the woman I was originally in contact with and a partner of the firm. When I sat down, he told me that despite receiving my letter, and all other documents, he wanted to hear in my own words what happened—as much as I could remember. I proceeded to tell the story of my rape, yet again. It was still difficult to discuss, but it was getting easier, not because I had it memorized, but because the therapy I was doing helped me to separate the physical reactions to the trauma from the words as I was telling it. There were still a few moments of tears, and not just from me. I will not use her full name, but the attorney I originally spoke to "N," also cried at times. It was not as obvious as my tears, but every so often when I would talk about certain things pertaining to the rape, I would see tears in her eyes as well.

When I finished telling them what had happened, the firm partner told me they would be more than happy to represent me, that he would have no problem putting me on the stand; I was a very credible witness. He then began to explain a few things that I already knew would eventually happen. I would have to be evaluated by their psychiatrists, that my entire past would be under scrutiny, specifically my sexual history; who I had sex with, how many people I had sex with, what, if any, fetishes or kinks I may have had, if I had ever used heavy drugs, and lastly, the expenses that would be incurred by me, and the fact that the case could take years to settle.

I told them both that I was fully prepared for these things to happen. I knew that my being in counseling would be flipped around to make me look "crazy," that I would be made out to be liar, a jilted lover, and the Whore of Babylon. I understood it all and told them I was still willing to go through with everything. I remember that at this point, "N" sat back in her chair, rubbed her forehead, and looked down at the table as the partner spoke. He said to me, "There is one more thing you need to know about. Let's say that this goes through the court, you have a trial, and you win the suit, I am guessing you would want some monetary compensation, correct?" I told him that I would, and at minimum would want my expenses paid for. I quickly looked over at "N" and saw that once again, there were tears in her eyes. "The thing is, even if you were to be awarded monetary compensation, it could be years before you see that money."

I told him I understood this as well, especially depending on the amount that was awarded. Then came the big punch in the gut ... "The reason it could take so long to see any of that money is because after the decision is made, both of those men could countersue you for defamation of character." I sat there, stunned, and not knowing what to say. All I could think was *what the fuck is actually happening right now?* He told me that I had a couple of weeks before I had to decide, but I was suddenly overcome with a different anger than I had felt from the beginning of the entire process. I am sure it registered on my face because he asked me if I decided not to go the route of the civil suit, would I want anything else to help bring me closure? "I would like them both to apologize to me, here, at this table in front of everyone in this room, my family and their wives."

He told me we could make that happen, and then asked me, "If we do make this happen, would you be willing to sign a non-disclosure agreement?"

"Does that mean I can never speak of it again?"

"Yes. That is exactly what that means."

"Then no. I am not willing to sign that. I was silent for many years through no fault of my own, and I will not be quiet about it now."

I continued to tell him that I would not need the time to decide about filing the suit, because I would not be filing one. They had already violated me once, and they would never get that opportunity again. They did not get to sue me for defamation when I knew that what they had done was far worse than them feeling like I was attacking their character. "N" had been a federal prosecutor at one point in her career, and the partner had asked me if I remembered being taken over the state lines at any point during the time, I was with them. Unfortunately, the answer was no, but if they had, he would have been able to file criminal charges on a federal level because there was no statute of limitations on rape for the federal courts. He asked "N" to look into anything else that could be linked to the federal level and for me to call him in a week to see what, if anything, they came up with or to let them know if I decided to go ahead with the suit.

I left the office, feeling those tears starting to well up again, but I did not let them fall until I got in my car. It was a quiet ride home and after I dropped my sister off at her house, I went back to my mom and dad's house where I was staying. I spent the rest of my time in St. Louis visiting with friends. When I returned to Virginia, I went to my therapist. We talked for a while longer than normal one day, and by the end of the session, we had talked a lot about justice and what it looked like to me. One path to my "getting justice" or moving toward closure was to start a group therapy setting or support group for anyone who needed to bounce things off of others, especially survivors of trauma. Tina was more than happy to help get that started and within two weeks, we had our first group meeting.

This all felt like a very good start, but somehow it just wasn't enough. I was still very angry, very hurt, and not handling any of it well. I was still taking my anger and frustration out on everyone around me, mostly my husband at the time, David. I still went to therapy, and I always attended the group sessions, but the anger fueled me, and I needed to let it go. It was making me weak and bitter, and this wasn't who I was. During one of my sessions with my therapist, she asked if I was still journaling. I was but not like I used to. She suggested I be more consistent with it as well as looking into taking some medication for anxiety. I promised I would talk to my doctor about it but wanted to switch doctors. One of the ladies in the group, Jacquie, gave me the name of her doctor so I contacted him. Within the week, I was on a medication for anxiety and depression.

One night, I woke up in the early morning hours of the day, once again crying from nightmares. I got out of bed as quietly as possible and let the dogs outside. After I let them back in and grabbed a blank notebook and pen, I sat on the couch with all three puppies curled up with me on a blanket. I put pen to paper in the early hours of the morning ...

"Kris woke in a sweat, tears dried on her face ..."

It was then, in those early hours, in a quiet house with my fur babies curled up next to me, I took the first steps toward taking back my power.

Sharon at work with the medical examiner working on reports.

Repelling down a steep hill toward a scene.

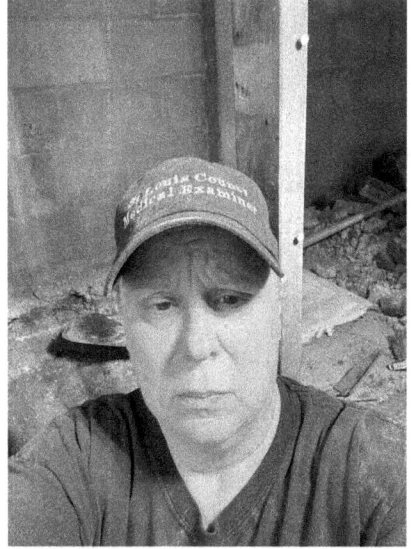

Sharon working to uncover remains.

First book release/signing event.

MARKOS PAPADATOS

"You can't put a limit on anything. The more you dream, the farther you get."

– Michael Phelps
Olympic gold medalist swimmer

Beyond "The Bay"

Back in December of 2020, I was pitched a big, interesting story by a publicist friend of mine. At the time, I turned it down and said, "absolutely not," thinking it would be too risky for me and for my readers. The topic was about covering "fair, accurate, and inclusive LGBTQ representations in the media and the impact of those projects."

Up until that point, this issue was something I had never covered before, and I thought that it could affect my reputation since it would stand out in a largely uncontroversial body of work covering music, entertainment, sports, magic, and lifestyle feature articles. Also, I was preparing for some big events in my personal life, so I was too busy working on arranging those.

When the events in my personal life fell through, a few months after that, I decided to give that pitch a second chance. I was always a huge fan of the 23-time Emmy award-winning digital drama series "The Bay," created by Gregori J. Martin, on Amazon Prime Video and Popstar! TV. I thought that if I were to cover this, I would use my favorite show as a vehicle to tell this moving story, and I did.

They filmed their sixth season at that point, during the quarantine, thus becoming one of the first series to resume taping during the pandemic. The show was about to tell its first same-sex marriage storyline featuring the dynamic performances of Mike Manning and Eric Nelsen as Caleb McKinnon and Daniel Garrett respectively. Both actors scored Emmy nominations for their performances, with Mike Manning triumphing with the 2021 Emmy trophy.

I knew that this was my chance to do what I did best as a journalist, so I did extensive research (on LGBTQ stats during the COVID-19 pandemic), thorough interviews with the filmmakers and actors, and wrote the piece as a storytelling article, where my goal was to educate, inform, and entertain my audience, all in one.

This article titled "The Bay actors earn Emmy recognition for same-sex marriage storyline, filmed during the pandemic," turned out to be my biggest and most beloved article (and probably one of the longest, spanning 12 pages) published in 2021. The response I received globally was overwhelming, in the best way possible. My friends and peers in the industry were proclaiming it one of my "career-best" pieces up until that point, and looking back, they are absolutely correct.

It was an important story that needed to be told, especially at a time when most of the news on TV is negative, and I was honored to have done it on "The Bay." Many thanks to showrunner Gregori J. Martin, executive producer and writer Wendy Riche, as well as Emmy winners Kristos Andrews, Mike Manning, Eric Nelsen, and Emmy nominee Gregory Zarian for providing me with great and informative quotes for this article.

In addition to the praise, warm comments, and positive feedback from readers above, this profound article was my 2021 entry for the "Outstanding Journalism Article" for the GLAAD Media Award, as well as my 2021 Pulitzer Prize submission for "Outstanding Feature Article."

All in all, this article did stand out in a prolific, heartwarming, and positive manner.

Sometimes we live in fear of taking a new path or what people will think if we try something new; however, there is power in choosing to take the next step in your own journey and evolution, without fear.

I hope others find inspiration in this story, and the other stories in this new book, and I hope it helps them find themselves and their own voice along the way.

"The Bay" actors earn Emmy recognition for same-sex marriage storyline, filmed during the pandemic

By Markos Papadatos

Mike Manning and Eric Nelsen in "The Bay" on Popstar! TV. Photo Credit: LANY Entertainment.

If you have ever read English Author Jane Austen ("Pride and Prejudice," "Sense and Sensibility," and "Emma"), you might remember dissecting her literary works line by line in an effort to discover all of the hidden symbolism and metaphors. You find the common ties that bind the story together and radiate an important message to the reader.

Austen's verbiage suddenly becomes more than just words on paper but rather something intriguing and compelling that takes readers on both a personal and a literary journey. The words and their syntax allow one to vicariously live through the author and become an active participant in the story.

Of course, the digital drama world has the potential to draw all the same parallels with the added benefits of our visual and auditory senses. That was what the 23-time Emmy award-winning digital series "The Bay," created by showrunner Gregori J. Martin, was able to accomplish in its sixth (and most recent) season with its first-ever same-sex marriage storyline, that was set during the COVID-19 pandemic and helmed by actors Eric Nelsen and Mike Manning.

American entrepreneur Sam Altman once said: "Young people willing to push super hard to make something happen are among the most powerful forces in the world." Actors Eric Nelsen and Mike Manning of the Emmy award-winning series "The Bay" are such inspiring young individuals, actors, and storytellers.

Eric Nelsen at the 48th Daytime Emmy Awards at the ATI studios on July 18, 2021, in Burbank, California.
Photo Courtesy of NATAS, Daytime Emmys.

Nelsen and Manning both earned 2021 Daytime Emmy nominations in the category for "Outstanding Performance by a Supporting Actor in a Daytime Fiction Program" for their captivating performances as Daniel Garrett and Caleb McKinnon, a same-sex couple in "The Bay."

Caleb McKinnon is a new school teacher in Bay City, and Daniel Garrett is the cousin of "The Bay" protagonist Pete Garrett, who is portrayed by Emmy winner Kristos Andrews. Daniel is a character who a lot of people can relate to because he consistently finds himself at the wrong place at the wrong time for all of the wrong reasons.

Daniel comes to a startling realization that he needs to be true to himself, all while he continues to struggle with the consequences of his sexuality.

Their characters were old friends from the past, but when they reunited, their bond and love for each other grew stronger than ever as they came to terms with their sexual orientation. They moved in together, fostered their relationship and love, and ultimately, tied the knot as part of the show's first-ever same-sex marriage. They were presented with a conflict, where Daniel's former love interest Matthew Johnson (played by 2021 Daytime Emmy nominee Randy Wayne) is back in the picture and expressed his feelings for him.

Randy Wayne, who plays Matthew Johnson, in "The Bay" Season 6. Photo Credit: LANY Entertainment.

"The wedding storyline was incredible because you finally see my character, Daniel, the happiest he has ever been," Nelsen said. "Daniel has struggled so much over the last few seasons. He has endured heartache and coming to terms with who he is and with acceptance, from not only himself but the world around him."

"Finally, we see him fall in love with Caleb and that love is reciprocated, and for it to end with this beautiful marriage was a magical moment for me because it got to come full circle for Daniel's journey. Where he started was so drastically different from where we see him at the wedding. It really was beautiful, not only from a performance standpoint but from a personal standpoint to complete that journey. I am so happy that we got to make that happen," Nelsen elaborated.

Actors Najee De-Tiege, Eric Nelsen, and Mike Manning in "The Bay"
Season 6 same-sex wedding scene.
Photo Credit: LANY Entertainment.

Particularly impressive was the fact that this poignant storyline, and the entire sixth season of the hit digital series "The Bay," was filmed under strict COVID-19 protocols while quarantining in a massive ranch in Santa Ynez, California.

Nelsen and Manning were nominated for the Daytime Emmy alongside such veteran actors as Tristan Rogers ("Studio City"), Neil Crone ("Endlings"), and Cheyenne Jackson ("Julie and the Phantoms"). On July 18, Manning bested the stiff competition and was proclaimed the winner, thus taking home the coveted Daytime Emmy Award.

Mike Manning winning at the 48th Daytime Emmy Awards.
Photo Courtesy of NATAS, Daytime Emmys.

Manning, who is openly bisexual, was hailed by Digital Journal for being "Emmy-worthy" for his dynamic portrayal of Caleb that truly ran the gamut. Caleb was honest and the voice of reason when Daniel needed it the most, as well as his pillar of comfort, when Daniel discovered that his former lover, Matthew was battling for his life against COVID-19.

This was a timely story in and of itself, which illuminated the direct and indirect effects of the Coronavirus on LGBTQ people and those who love them, at a time when a Human Rights Campaign (HRC) and PSB Insights study found members of the LGBTQ community are more resistant than the general population to the vaccine effort.

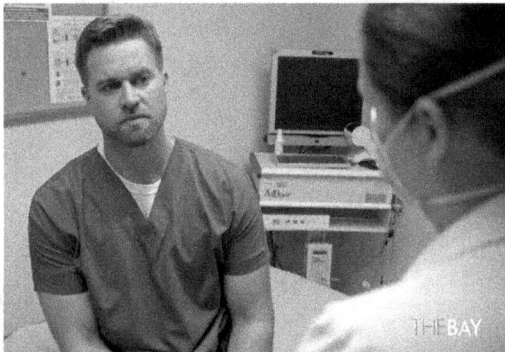

The character Matthew Johnson (played by Randy Wayne) learning about his COVID-19 diagnosis on "The Bay" Season 6.
Photo Credit: LANY Entertainment.

In Season 6, Nelsen was able to extend his acting muscles even further than the previous seasons in a subtle, raw, and unflinching fashion as the world was plagued with various social and cultural issues. Nelsen delivered an authentic and brave acting performance that was described accordingly as a "quiet storm," where he conveyed a wide spectrum of emotions, that hit viewers like a shot in the heart. They both deserve to be commended for accurately portraying the loving relationship between a same-sex couple.

"I thought that Gregori gave us a masterpiece storyline to work with," Nelsen said. "Although we are at the height of our love story, there is still a tragedy that gets thrown in there. My character finds out about the death of his former boyfriend all while Daniel is madly in love with Caleb. To play these opposing stories was difficult but also so rewarding to see it all come together on TV, and it worked really well. Getting to work with actors such as Mike Manning and Randy Wayne heightened my journey as an actor and an individual because they are giving so much, as well as they are taking."

"I am so happy and proud of Mike [Manning] for winning the Emmy Award this year, and I told him that if it wasn't me, I would pray it would be him, and I really meant that, I am so happy that it was him. I am sure this is just the beginning for both of us, I don't see this storyline stopping anytime soon," Nelsen added.

Eric Nelsen and Mike Manning in "The Bay" on Popstar! TV.
Photo Credit: LANY Entertainment.

The same-sex marriage storyline in Season 6 was written in a bold, original, and refreshing manner, and it stood out to Manning for several reasons. "The thing I appreciated most about the way the LGBTQ storyline was written was that it was simply 'about the love' between the two characters," Manning said.

Executive producer, writer, and executive story editor Wendy Riche stated, "When Gregori and I write, we always think about the emotional honesty of each character. If we are true to them, then the audience will be able to relate and take the journey with them. Love is not defined by gender or sexuality. Love is love. That is what we hoped to bring the viewers in Season 6 of 'The Bay.'"

Executive producers Gregori J. Martin and Wendy Riche. Photo by LANY Entertainment.

Creator, Head Writer, Director, and Executive Producer Gregori J. Martin shared that he is "hoping Daniel and Caleb's love story had and will continue to have a major impact on viewers, in the sense where more folks around the world, by watching the series, understand that love is love, no matter what gender they identify as."

"When I wrote the Daniel and Caleb love story, I didn't write an LGBTQ storyline," Martin said. "I wrote a true honest love affair between two men, and it's one of my favorite stories to write. Eric Nelsen and Mike Manning are always so authentic in their delivery as childhood best friends who fall in love. It's absolutely beautiful to see these two very talented Emmy-winning actors bring their characters to life so flawlessly."

Martin continued, "It's an honest story of two young men, who once had a forbidden attraction toward one another, now be able to break free of their chains built on fear and finally live a life together. Will there be bumps in their road ahead? Certainly, or you wouldn't keep the audience entertained, but if watching Daniel and Caleb's love story helps any viewer break their own chains of living a life of fear, or growing up with a primitive mindset, then I've done my part, if only a little, in making this world a better place, routed in honest true love, and not false judgment."

"The writing didn't play into the (sometimes harmful) LGBTQ stereotypes," Manning said. "The story could have been about any relationship, and it just so happened to be about two guys. You saw these people move in together, get closer, and then get married. It was about celebrating love, family, and commitment. I think that's important for people to see, that this relationship is just as 'normal' as any other relationship."

Manning felt extremely privileged to be nominated in that Emmy category along with his luminous on-screen acting partner Eric Nelsen, who is an Emmy award-winning actor and producer in his own right. "I am really grateful that I got to work with Eric on this particular storyline in Season 6. He's an incredibly talented actor and I think between the two of us, we were able to bring authenticity to the relationship. We joked before the Emmys, 'if I don't get it, I hope you do,'" he said.

"There is nobody I would rather work with like that on the show … except maybe Dave Franco," Manning said with a sweet laugh.

This same-sex wedding storyline (and the Emmy recognition for Nelsen and Manning) is so much bigger than both of the actors involved. The storyline has made a substantial overall impression on viewers, fans, and it will certainly impact future generations in terms of accurate and inclusive LGBTQ representations.

Mike Manning holding his Emmy trophy at the 48th Daytime Emmy Awards. Photo Courtesy of NATAS, Daytime Emmys.

Manning remarked, "I said this during my acceptance speech at the Emmys: 'If there is one young person at home who watched Season 6 and realizes that they are capable of the type of love that these characters find and are celebrating, then the entire season for me was worth it.' Visibility and representation matter."

"Storylines like these on TV matter," Manning admitted. "I am glad that the writers (Gregori J. Martin and Wendy Riche) gave Eric and me such an important storyline that I hope will help anyone in the LGBTQ+ community struggling with self-acceptance or self-love."

Executive Producer, Co-director, and Lead Actor of "The Bay" Kristos Andrews expressed, "Acting is the art of being human. We are each greater and more expansive than life tends to condition us to believe."

"There is an innate value to tap into within this craft, and whether it be pleasure, pain, or experiences beyond what's verbally describable, finding the meaningful silence within the work begets a greater acceptance towards life, understanding toward others, and growth of our consciousness," Andrews added.

"Love is love," exclaimed Gregory Zarian, Emmy-nominated actor of the digital drama series "Venice: The Series." "When told truthfully and honestly, race, color, shape, size, and sex, do not matter."

Aside from the same-sex marriage, "The Bay" touched upon such relevant social issues as the COVID-19 pandemic, panic buying, police brutality, and the "Black Lives Matter" (BLM) movement that were simultaneously affecting our country.

Eric Nelsen as Daniel Garrett in Season 6 of "The Bay."
Photo Credit: LANY Entertainment.

Season 6 of "The Bay" was an acting and filmmaking work of art, which is a testament to the four Daytime Emmy Award wins (and 13 nominations in total) that it received in addition to their other previous wins and nominations that now total 23 Daytime Emmy wins and 56 nominations. "My heart is full right now for my beautiful and talented cast on their Emmy wins in the daytime fiction program categories. So well deserved and I love you all very much. Keep killing it," expressed Gregori J. Martin.

This same-sex relationship storyline, in particular, was able to take people through multiple layers and levels that many digital series could only dream of. It was able to take viewers on a personal journey that was bumpy, raw, conflicting, exhilarating, and explosive.

The story further met the cultural moment at a time when GLAAD's annual "Where We Are on TV" report found the representation of LGBTQ characters on television (including streaming shows) to be shrinking, down from 10.2 percent to 9.1 percent. Stories like the one scripted by Gregori J. Martin and the writing team for Nelsen and Manning, offered not just representation, but positive representation to their global audience, and it served as a powerful response to that unfortunate trend.

Eric Nelsen and Mike Manning in "The Bay."
Photo Courtesy of LANY Entertainment.

In the end, Eric Nelsen and Mike Manning tackled the storyline's subject matter with much delicacy, authenticity, and sentimentalism. Their vulnerabilities as actors and performers were the audience's reward since they were able to connect directly to the hearts of people. "I think LGBTQ characters on television are still more sparse than they should be," Nelsen acknowledged.

"To be a part of a journey of an individual who shows that struggle is real and that they are not alone meant a lot to me. I knew it was a part of a bigger picture, and to this day, there have been a lot of people writing to me that my storyline helped them in more ways than I will ever know. That alone makes it so much more worth it for me as an actor, especially to know that I can touch people who might be struggling with the same issues. When they see somebody else battle it and come through on the other side in such a beautiful way is quite inspiring," Nelsen explained.

"I hope it inspires other people who struggle like Daniel to find their path and their light, and to know that it's okay and that they will be accepted and loved. The beauty of showing this on television is that it is going to reach people, and that's the most gratitude that I get from it," Nelsen further added.

Most importantly, this same-sex marriage made a statement and reminded us that even if darkness and adversity are surrounding you, there is always light and hope within reach. It was like a trance that nobody wanted to come out of. Take that Jane Austen.

About Markos

Markos Papadatos is an award-winning editor, journalist, and educator from Long Island New York. He has authored over 20,000 articles over the past 17 years. He has interviewed some of the biggest names in music, entertainment, lifestyle, magic, and sports. He averages anywhere from six to 10 interviews a day (and receives an average of 2,000 pitches daily). He has been published in two languages (Modern Greek and English). He serves as an associate producer of "The Donna Drake Show," which airs on CBS New York.

A bestselling author and poet, Papadatos is a seven-time consecutive "Best of Long Island" winner, earning 15 "Best of Long Island" titles, and in the past three years, he was recognized as the "Best Long Island Personality" in Arts & Entertainment, an honor that has gone to Billy Joel six times.

He was honored as "Journalist of the Year" twice by the Greek-American Press in 2017 and 2021 respectively.

Thanks to his fast turnaround rate, the biggest and most powerful publicists in the industry have proclaimed him "the fastest pen in the USA."

MARY MARKHAM

*"Faith is taking the first step,
even when you don't
see the whole staircase."*

– Martin Luther King, Jr.

Empowered By Faith

Are you living your best life? Free of fear? Or lost hope? Unable to take that leap of faith because you're stuck in life's quicksand, fearful of the unknown, anxiety from judgment, playing the compare/despair game and feeling unworthy, while trust issues are making you sink even deeper, with no hope of freedom?

Hopeless! Not only did I feel hopeless, but my self-worth, self-esteem, having no voice, and what seemed like everything else in between were keeping me stuck in life's quicksand for years. I began sinking into darkness while going through the motions—whatever that looked like at the time. No longer having a plan or desire to fulfill dreams, fear kept me from speaking up about anything. Fearful of hearing again, "You're not intelligent enough or good enough," or "Women are better off if they are just silent" or continually being interrupted when I did try to speak. The darkness of hopelessness seemed endless.

Why dream? Only to wait for someone else to come into my life and sabotage my dreams anyway, or steal my idea, or tell me:

"You're crazy, it'll never happen; someone else does that or did that and it didn't work."

"You're not good enough!"

"You don't have enough money."

"You're not smart enough, so why try?"

The continuous lies and comments were someone else's story, but I believed it to be my story. My light was so dim, and the trust issues were so deep that it kept me believing my dim light would stay dim forever.

The shame/blame game held me back and affected me more than I realized. My days were wasted trying to understand the "why," and the dark secret of being sexually abused at 14 years old kept me hidden from shining my light. I felt stuck: physically, mentally, emotionally, and spiritually for too many years. A secret and shame that kept me from believing I could ever do anything right or trust anyone again. My negative thoughts, words, and actions pulled me into the opposite direction from dreaming, having aspirations, or experiencing happiness and joy in my life.

How do I find that young, vibrant girl inside, who once smiled at everyone, helped anyone in need, and before the abuse, sang "I'm on Top of the World" as I walked to school.

Have you ever felt like you lost your own identity?

My daily life became a continuous high and low. Sinking ever so slowly when I believed the highs were temporary, I intentionally tried to never get too excited as my lows kept me from feeling lit up. The lows could have ended my life if I would have let them. But something deep inside was stronger and kept my dim light from ever going out. I longed for that fairytale life: happily married forever, living in a beautiful home, and having children to fill my home with love and laughter. Although my dream did come true: marriage, a beautiful home, and wonderful children, it was the real authentic person inside who was running away and feared showing herself. My inner-child self was afraid to live life to its fullest and afraid to shine. And when my lows caused my light to flicker, I surely thought it would burn out, especially when my fairytale life headed down the path of divorce. I was broken, but something deep inside told me not to give up. *What was that something?*

It was 3 a.m. when I woke up crying, praying, and questioning everything in my life, wondering, what am I going to do as a single mom of two teenagers? I journaled every thought, good or bad, to help clear my mind. That little voice inside of me started talking, and I was finally listening. My prayers became conversations with God. I was scared, felt alone, and wanted to know how I could ever trust anyone again? It was then, my aha moment, when I began reflecting on my life and all the times I felt challenges, struggles, and defeat. I asked myself, *What was that "THING" that helped me get over it, past it, or through it?*

Deep dive! I dug so deep inside and finally realized that through every challenge or obstacle that came my way, the one thing I never gave up on was my faith. When I felt hopeless, my faith was still there. I had just closed my mind and my eyes believing it didn't exist anymore. My shame, anger, and frustration, darkened my hope and put my faith on a shelf just like an inspiring book that I'd read when I had time. *But when is there time? Will I ever have enough time to turn up the dimmer switch on my faith light? What do I need to do differently to brighten my light? Have a voice? Smile? Feel alive again?*

As long as I kept looking at the negative, the past, or the things that were going wrong, I stayed in this darkness—a very dim-lit life. I needed to heal internally before I could slide that dimmer switch up and shine brighter. But how?

What was I doing wrong?

What was I not doing enough of?

How could I dig any deeper to find my answers?

So … I thought! The deeper my reflections became, the more I realized how my faith got me through those challenging times. After the abuse at 14, God put an angel in my life to give me hope, love in my heart, and a belief that my dim light would one day shine brighter than I could ever imagine. I needed to believe, keep the faith, and trust God and the process.

When I suffered a total of four miscarriages, I was blessed with nurses who held my hand, hugged me, and gave me the support I needed. I had an amazing doctor who repeatedly reassured me to have faith and not give up. My faith gave me two beautiful living children. My faith allowed me strength when I could have hit rock bottom—spiraling downward with no desire to reach higher. During this time, I was surrounded by my sister-in-laws who all had healthy deliveries for each miscarriage that I suffered. This "THING" called faith allowed me to never give up!

What is faith? Is it a religion?

Faith's definition:

1. "Strong belief or trust I have faith in our leaders."

2. "Belief in God."

3. "A system of religious beliefs; religion people of all faiths."

Faith is having complete trust or confidence in someone or something; a strong belief in God or the doctrines of a religion.

For me, faith was never attached to a specific religion. It was believing in the unknown, unseen, and allowing the process to unfold in God's way and in His timing. When I realized that I was no longer breaking down but breaking through and no longer falling apart but falling into place, I felt empowered to be the person I was created to be and an example for my children.

Everyone's level of faith is different. When my son was two years old, while in the laundry room, he managed to pull a chair over to the fish tank and scoop out his little friends. When I became aware of what took place, his face was priceless yet fearful. While trying my best to stay calm, I repeatedly explained that the fish needed water. With his bottom lip quivering, in his quiet little voice, he said, "I … I … just wanted them to be my friends." We said a little prayer and gently put them back in the water and watched them swim around, thankful for the gift of water. My instinct was for the fish to survive by putting them back into their environment, not assume they would just die because they were out of their water. I had faith. My faith kept me strong, and the dimmer switch was slowly and steadily sliding upward keeping my light shining brighter.

I realized how bright my faith was shining when I saw a mob of teenagers preparing to attack a young girl. Stepping into the middle, without fear, I managed to break it up without violence. And days following that incident, I saw a young girl, whose boyfriend drove recklessly into a retention area and left her alone to take the heat. She was clearly upset and questioned why someone who said he loved her would leave her so easily? Without even the slightest hesitation, I asked her if she believed in God and had faith? She replied, "Yes!" I continued to shine my faith light and explained how that boy showed his selfish side; this wasn't love. It was as though she processed this new information: How could someone who said he loved me treat me this way? I explained how love should never come in the form of abuse, whether mentally, physically, emotionally, or spiritually. I felt called to share, shine, and warn her so she too could have the faith to get out of this abusive relationship and be treated with love.

Six months later, she knocked on my door and shared how she was grounded and had broken up with that boyfriend. She began focusing on her new faith journey. God had a plan for both of us to meet at that very moment and time. His plan is always bigger and brighter than we could ever imagine.

I had the strong urge to connect the dots, physically, mentally, emotionally, and spiritually and reflect on how each person and situation along my life's path were there, happened for a reason, allowed my faith to deepen, and brought that vibrant attitude and smile back to my face. Each experience allowed me to learn, teach, or set new boundaries in order to surrender to what is, let go of what was, and have faith in what will be.

While discovering one's faith means something different for every person, my faith sees the invisible, believes the incredible, and receives the impossible. Faith is having complete confidence, without questioning belief, in the truth, value, or trustworthiness in a person, idea, or thing. I believe because of my faith I have never given up hope or the belief in myself, even if at any time when it felt like I was heading down that path. I suppressed my faith and rediscovered it when I was ready to embrace it. I needed to change my mindset and let go of the old and others people's stories that did not serve me. The more I let go, the deeper my faith became. It empowered me with strength, confidence, and light when it needed to shine because I was growing physically, mentally, emotionally, and spiritually.

Physically, I discovered the importance of being present—to physically live in the moment. I didn't allow myself to sit in the mindset of my past or focus on the disappointments, frustrations, or the "whys" or worries about the next minutes, days, months, or years ahead of me that I had no control over.

Mentally changing my mindset and becoming intentional about my time alone, enjoying being, and doing whatever it is at any given moment was a process. Having this mindset shift allowed me to embrace any and all emotions that came forward. It's not that I ignore or don't experience emotions, because life still happens and unexpected things can challenge me if I allow them to. However, I have learned to become more aware of my feelings, embracing them, and staying present by enjoying each moment I am given.

These steps allow me to grow spiritually as well. I have deepened my faith and discovered how to shine by sharing my spiritual gifts. I no longer compare myself to others by embracing my authentic self: to be present, trust God, and the process. I no longer worry about the unknown, but I embrace every new opportunity with thanksgiving and gratitude.

My deepened faith connects me spiritually to my inner core and having complete trust in something higher than myself—God. I have gained understanding, wisdom, and knowledge with an abundance of strength to walk down life's path with a brighter FAITH shield protecting me from challenges, obstacles, and what I call the Devil's darts. Life is a choice, and I choose to look for the good in every situation; be thankful and grateful for everything and everyone; embrace my emotions and every moment by being present in life; the true gift that it is.

When one door closes another door does truly open. Whether the door was meant to open or close for me personally or for someone along my path, having faith through my thoughts, words, and actions has given me the encouragement to keep moving forward with a new positive mindset.

I continued to connect the dots of my empowered faith and how I was the "world" to that young girl so many years ago, whose boyfriend drove into the retention area and left her there alone. That experience could have taken her down a negative path chasing after a life of loneliness and abuse but sharing my faith empowered her to turn her life around for the better.

Or the time I showed no fear when I heard a loud noise, "BANG! BANG!" It was a chain hitting metal which came from the park across the street from where I lived. He was tall, slim, had a few tattoos, and wore baggy jeans and a T-shirt. He was obviously a distraught and angry teenager. He stayed focused, staring at the center of the slide as he whipped a huge chain on the slide of the playground equipment. I somehow felt his pain. Without hesitation, I compassionately walked over and empathetically said, "I can see you're upset about something, but is it necessary to destroy the children's equipment because of something hurting inside of you?" He didn't say a word but slowed his aggression of force with each new whip. I then asked him if he would sit with me at the picnic table. The whipping stopped as he stared at the equipment; he contemplated what to do next, sit, or just walk away?

I asked him again, "Are you okay? Do you want to talk about it?"

Without looking at me, he pulled the chain closer to his body, turned around, and sat at the picnic table. I asked again, "Honey, would you like to talk about what's upsetting you?" I sat, waited, and listened.

He said, "Why do you care?! All my life, I've been told how 'bad' I am and how no one cares about me, so why do you care?"

I said, "Everything God makes is 'GOOD,' and you are no different! God made you, and therefore you are good."

Faith empowered him to share his broken stories. We talked, and I reiterated what a good person he was. I saw tear-filled eyes as he stood up. He walked around the table and asked if he could give me a hug.

He said, "No one has ever cared or listened to me like that before. Thank you!"

I said, "God will give you the strength to get through your pain; just have faith." He calmly left with a new purpose.

That situation could have gone very differently had I approached him with demands and a negative attitude. My Faith empowered me to stay strong, trust God, and believe in the process. The list of people who have walked into my life, opened doors to allow my light to shine, or closed doors that no longer served me well, is forever long. The many influencers throughout my life have helped me to be a good mom, friend, teacher, and mentor. Each year I grow deeper in my faith, learn to let go of the things I have no control over, and live intentionally in the present. My life changed. I've changed. I have a voice, confidence, compassion, and empathy to help others along their paths. My empowered faith has opened doors to lead others through my Spiritual Life Coaching and make a difference by sharing my authentic self, worthy of more than I could ever imagine, and fulfilling dreams I never thought possible. With God ALL things are Possible!

How did I take back my power?

Discovering my true authentic self has allowed me to connect the dots of my life. The many platforms as a teacher, mentor, author, coach, speaker, and podcaster allowed me to turn up the dimmer switch, shine my own light, and acknowledge my voice. My deepened faith empowers me to dig deeper in understanding, awareness, and to serve my purpose doing what I am called to do.

My belief is that we are all called to teach, learn, or share our spiritual gifts to help empower others. I either learn, teach, or share something with each person who walks into my life. Each experience leads me to build myself into a stronger person, heal myself as I walk along my path, and thrive and shine as a spiritual life coach. As a coach, my empowered faith helps me to help others get spiritually connected, feel valued, heard, and transition through midlife changes, in order to build, transform, and thrive by serving their purpose.

Each day is a gift for me to witness how my own faith empowers others to deepen their understanding and increase positivity and self-awareness.

"To the world you may be one person but to one person you may be the world." – Dr. Seuss

Have you ever thought about the difference you make in a person's life? That one small gesture or comment you made may have saved their life? Or maybe someone saved yours?

Believe and have FAITH; it just might change your life more than you could ever imagine!

About Mary

Mary is an author, co-author, ICF-certified spiritual life coach, speaker, podcast co-host and soon to be Spiritual Director.

Mary is the author of *Seeds of Life*, her first children's book and *In God's Hands, a Memoir of Hope*. She is also a co-author of *Manifesting Your Dreams* and *Miracle Effect*.

As a Spiritual Life Coach, Mary guides others along their journey of personal and professional growth. She helps those clients, who are ready to let go of fear, get unstuck, take a leap of F.A.I.T.H., discover their purpose, and embrace new opportunities with clarity and confidence. She has spent decades teaching, mentoring, and leading others. She provides a wide range of programs and services, which includes a healing road map, mentorship program, individual coaching, and workshops. Please visit maryamarkham.com to learn more.

Mary loves to travel and spend time with her husband, Craig, and their beautiful blended family and grandchildren.

LinkedIn: https://www.linkedin.com/in/mary-markham

Instagram: maryamarkhamlifecoach

FB: Mary A Markham Spiritual Life Coaching

Email: mary@maryamarkham.com

NATALIE M. MILLER

"Each one of us has a fire in our heart for something. It's our goal in life to find it and keep it lit."

– Mary Lou Retton

From Powerless to Powerful

It was September 2021, and I was working remotely in the legal department at Harley-Davidson since the pandemic upended our lives in March, 2020. My employment contract was scheduled to conclude at the end of 2021, so I was actively seeking new employment and speaking with recruiters—determining what I wanted to do next. I was ready for a new challenge, but I was interested in so many potential new career paths that I didn't know which one to focus on. I researched what career paths I was drawn to, the educational requirements, the skillset needed, as well as the salary information and opportunities for growth. I registered with various employment and recruiting companies which led to me receiving a lot of hits from recruiters. I was recruited for contract administrator positions, paralegal/legal opportunities, and high-level administrative roles. I had become well-adjusted to working fully remote, and I really enjoyed the flexibility. I was hoping to secure a remote position, especially since that work trend was here to stay.

After having been recruited for multiple positions, only a few sparked my interest. I was especially enlightened when I received an email about a remote administrative role, working for a web-based elder caregiver solution company. It was a unique opportunity, working for a feel-good company with a mission I felt could provide a better sense of job fulfillment. Additionally, the position paid surprisingly well; in fact, perhaps a little too well ... The email welcomed me to apply for the remote administrative assistant position, so I did. I received an email the next day, following my online interview, notifying me that I was offered the position. I was excited that I landed the role, but the entire process seemed way too easy, quick, and therefore, maybe too good to be true ...

Later that day, I received an official offer of employment on the company letterhead, as well as all the typical onboarding documents to complete. I was then asked to make a series of payments for office equipment and software, after advancing me a check, and instructed me to mobile deposit them into my account. That seemed odd to me, but the individual I was dealing with, assured me this was commonplace since the pandemic and remote work. After I executed two Venmo payments and a wire transfer totaling $6,700, for the office equipment, I received an alert that I had fallen victim to an employment offer scam. I immediately felt sick to my stomach and struck with despair, because this confirmed my worst fear—this was too good to be true. My head was spinning, my stomach was turning, and I couldn't believe what was unfolding. I wondered, *Why didn't I listen to my intuition when it was screaming at me? Why did I let my ego overrule my judgment?* I'm smarter than this, I knew this was suspect, and I can't believe I might be out $6,700! I felt completely powerless, like someone just knocked the wind out of me. I was blinded with overwhelming thoughts, despair, and regret. I couldn't even cry, so I decided to transfer my pain and anger into purpose and use it to take back my power.

My contract with Harley concluded on March 31, 2022. My team had a virtual going away get together and bought me a red (my favorite color) Harley-Davidson T-shirt and two coffee mugs. I was so grateful for the wonderful expressions of gratitude and goodbyes, the gifts, and for my years of working at Harley. I was ready to reflect on everything and determine what was next for me.

Fortunately, I recovered both the Venmo transactions after

notifying them of the fraud, so all that remained was the $4,300 wire. My bank attempted to recall the wire transaction, but the funds were already gone by the time it was requested. Unfortunately, my bank had no sympathy for my situation and was dismissive throughout the process. Since my bank failed to secure my signature to the wire transfer document before sending it through, I asserted the position that without my signature, the transaction was invalid. Therefore, the receiving bank/my bank should reverse the wire back to me. My bank's poor treatment toward me drove me to continue to fight to retrieve my money. I conducted research into the surge in scams since the pandemic and, specifically, those involving fraudulent wire transfers. I read that in most cases, banks reversed the transaction to their customer and those were fully executed transactions. The fact that my bank failed to secure my signature validating the transaction gave me even more leverage that it should have reversed the wire to me. After asserting my position to my bank and trying to work with them to reverse the wire transaction, they adamantly stated they would not be reversing it. As a woman of principle, I feel my bank should protect me and take ownership for their failure to conduct due diligence by recalling the wire when they realized they failed to secure my signature for the wire transfer. This combined with their dismissive and unsympathetic treatment, propelled me to take them on and pursue justice. I felt drop-kicked in the gut after going through this scam and even questioned my own judgment which was a difficult and painful experience. My mind was overwhelmed with thoughts and questions while simultaneously feeling so many unsettling emotions. I wanted to regain my balance with a sense of control and needed to take back my power; no matter the outcome, I felt the need to fight for myself.

As a discovery tool, I filed complaints against my bank to the Better Business Bureau, Office of the Comptroller of the Currency, and the Consumer Financial Protection Bureau. Since those agencies have no judicial authority, I filed a small claims action against my bank to recover the wire transfer.

Meanwhile, I continued to work with recruiters, field opportunities, and interview for multiple positions. I was intrigued when I was recruited for the position of business operations specialist with Caterpillar, Inc. Caterpillar is a solid company, and this was a new and different role, plus the position was mostly remote. I interviewed well with both the senior engineer and project manager for whom I would be working with and reporting to. I received an employment offer with Caterpillar as well as General Electric. I excitedly accepted the position with Caterpillar and would start work in August.

I was enjoying the summer, spending time with family and friends, and I tried to make the most of the next few weeks, before I started work at Caterpillar. I hung out with friends, had fun in the sun, attended concerts, went to the beach, and completed some domestic and passion projects.

On Monday, August 22, 2022, I had a court hearing for my small claims action against my bank. We had mediation a few weeks prior, at which time the mediator was able to get my bank to up its initial offer of $1,000 to $3,000. While I was grateful for this progress, I was standing firm in my position. I wanted to hold out and see how the court commissioner would rule, in hopes to recover the full $4,300. The court commissioner would have to consider the bank's Motion to Dismiss, my Response brief, as well as our testimony in court.

I appeared in person, and defense counsel appeared by video since he worked out of Pennsylvania. I'm generally a pretty confident, somewhat fearless woman and have only had a few moments in my life when I felt intimidated—this was definitely one of them. The judge saved our case for the end because he anticipated it would take more time due to its complexity. I had knots in my stomach since my arrival, and my nerves only heightened, as I watched the court room clear out.

The judge called my name; it was go-time. I walked up to the table in front of the judge, placed my court documents and color-coded notes on the table and sat down. The judge made his opening comments, and both defense counsel and I entered our appearances. The judge then asked me to begin reciting my case from the beginning. I instantly froze, and my mind went blank. I took a few deep breaths and began articulating my case and position. Then the judge asked defense counsel to state his side and position. This was a seasoned litigator working for a well-known law firm, and I was now going up against him in court.

Reality set in, and I suddenly thought, *What the heck was I thinking? I am in way over my head!* I couldn't help but watch the television in the corner of the court room while the defense counsel presented his case. Interestingly, he appeared noticeably nervous. I was surprised, but it did settle my nerves a bit. I wanted to interject and object to multiple things he said, but I knew I had to refrain from that.

We then proceeded to go back and forth, arguing our positions. I was able to successfully argue my points and rebuttals to defense counsel's defenses, except for one. I did not have an argument to refute this defense, so I just kept reiterating that the transaction was unexecuted and therefore invalid. I made the comparison to a check: if written out and dated, but unsigned, it is invalid and could not be cashed. Therefore, since the bank failed to secure my signature before sending the wire transfer through, it puts the burden on them to reverse the wire to me. The judge pressed me for a rebuttal to the defense's argument that negligence claims are preempted by the Uniform Commercial Code (UCC). I deflected and passionately reiterated my other points which the judge agreed with. However, because I did not have a rebuttal and this was law, he ruled in favor of the bank's Motion to Dismiss. I knew after my research, this was potentially going to be an issue raised in my case, but I still wanted to take the chance.

I felt crushed, like someone just sucked all the air out of my lungs. I was angry, but I had to accept the judge ruled in favor of the bank. Despite a few moments when I was frazzled, I feel I did pretty well conveying my position, while also allowing them to see the emotional toll the situation took on me. While I did not achieve the result I wanted, it didn't feel like a loss. I fought all the way to the end and pushed through my nerves and decided to take the chance and go in front of the court commissioner. The fact that I went toe-to-toe against a seasoned litigator and for the most part held my own, felt like a victory to me.

Even though I walked away with nothing when I could have accepted the $3,000 offer the bank extended at mediation, I don't regret powering through to the very end. I know if I would have accepted the $3,000 at mediation, I would have wondered if I could have achieved my goal if I went in front of the court commissioner.

Fighting through the nerves and uncomfortable emotions, while still being able to hold my own against a seasoned litigator, is an experience I'm proud to have gone through, despite the result. There wasn't much for me to process because I had no doubt I delivered the best case I could for myself. I was wrestling with the emotions I felt inside; however, I didn't have time to waste feeling disappointed. I had to mentally prepare for the next day which I knew would require all of my energy and focus.

I woke up on Tuesday, August 23, 2022, at 4 a.m. to get ready and travel to Chillicothe, Illinois for my first day at Caterpillar. When I negotiated my contract, we agreed I would commence coming on-site once a month and mainly work remotely. We agreed to revisit the arrangement after my first few months and discuss transitioning to working fully remote.

The drive to Chillicothe was nearly 3.5 hours, and it wasn't much of a scenic trip. I met with Bill, a senior engineer, on my first day. He showed me around the huge campus, provided me with my laptop, and coordinated trainings I needed to complete. He was a very nice, straight-shooter which I respected. Fortunately, I didn't have to put a full eight hours in and left early in the afternoon. I was mentally and emotionally exhausted from the previous day in court, first day nerves, and the long drive. I had to carry two large boxes with my laptop, monitor, and other hardware, so it wasn't a light load; plus I was wearing heels. The campus was huge, and all the hallways looked the same. All I wanted to do was get home and decompress. As I walked down the endless, empty hallways, somehow I got turned around. Suddenly, I was lost and all of my emotions caught up with me, combined with my lack of sleep; I almost broke down in tears. I felt momentarily powerless, but I pulled myself together, retraced my steps, and found my way to the parking lot and headed home.

For the next few months, I had a rigorous, nonstop schedule. My manager introduced me to engineers, inventors, and other team members. He trained me on various software applications, processes, led meetings I observed, and briefed me on projects I would be managing. He also registered me in a graduate course that was taught by one of the senior engineers I would work with, who was also an inventor.

Most days, I went from one meeting to another, then to class or a training session, without any time to digest all that I was rapidly learning. In my first week, my manager threw me into a meeting that I was forced to lead with engineers and inventors, on a process I had not even learned yet. My manager would give me a brief overview of each software application/process and expected me to be an expert in its entirety, in the blink of an eye. I was used to working in a fast-paced, challenging environment in corporate America, after having worked at Harley-Davidson for over 4 years, but this was insane!

I thrive on a challenge and love to learn, but I was getting thrown so much information simultaneously, with minimal guidance or direction, and no time to digest and understand everything. I never give up on myself or back down from a challenge, but it felt as if my manager was making it nearly impossible for me to succeed. When I asked follow-up questions on software programs or business processes, my manager only knew the basics which was about what I had learned in my first few weeks. I felt forced to learn in six weeks, what normally would haven taken three to six months.

He raised some big red flags—especially when he dumped all of his work on me and was barely available to answer my questions. Then he took credit for the work I had done! Another individual on the team made a point to say that he does not give any direction, and that is why nothing has ever moved forward. I was stuck doing what he had failed to do. It became clear my manager wasn't doing what he was supposed to for a long time; now he was pushing me to learn and digest everything in the blink of an eye. Then he wanted briefing on what he should have already known. He pulled some shady behavior, and I called him out on it. He tried to throw me under the bus claiming I missed a scheduling deadline, but in fact he cancelled the meeting. When I sent him a screenshot of the message, clearly proving he canceled the meeting, he did not apologize or take ownership and instead replied by saying, "Okay, no worries." Another time, he pushed me to follow-up with the status of one of his projects, for which he did not brief me on. However, when I asked him basic questions about it, he didn't know the answer. It all came to a head when my manager proved to me, he could not be trusted.

Meanwhile, my mother had fallen ill with COVID-19 in October, wound up in the hospital with pneumonia, and had a close call. This was especially concerning because I lived 3.5 hours away from her. I was mentally exhausted from work every day, so my weekends were spent decompressing from the intensity of my job. With the revelations of my manager's toxic behavior, I was now faced with the reality of my situation. I felt powerless. On top of the unreasonable expectations and intensity of the position, I could no longer trust my manager. I needed to reassess whether I wanted to continue in this role. My mother's health continued to decline, and she was scheduled for surgery in November at Rochester–Mayo Clinic. I notified my manager and his boss that I would be taking time off to be with my mom. I stated that I likely would not be returning and shared my concerns about my manager with his boss, so he could be addressed if/when I returned. Bill, the senior engineer who was well-respected and had been with the company for a long time, gave me a glowing response. He said he was really sorry to hear that, and I had become a valuable member of the team and would be missed. He further stated if I changed my mind in the next few days or if I wanted a position there in the future to let him know.

While I was packing for my trip to Rochester, I noticed an email from a recruiter about a position I submitted my resume for a few weeks prior. Between the stress of working for Caterpillar and the ups and downs with my mom's health condition, I almost missed the email. I didn't expect to hear anything back because I didn't have contract negotiation experience, but I remember thinking it would be a game-changer for me. I was elated to discover from the email that I was short-listed for an interview. The only time slot left for an interview was the next day at 11:30 a.m. which was during the time my mom would be in surgery. If it was any other position, I would have said I am not available because I have to be present at the hospital for my mom's surgery. However, this would be a life changing opportunity, so I told the recruiter I would make it work because I did not want to miss my chance.

The next morning, I met two of my mom's sisters, and we all took my mom to the hospital for her surgery. I then spent the next 2.5 hours preparing for my interview. Thankfully, I was able to secure a private room for my virtual interview. The closer I was to the interview and the conclusion of my mom's surgery, the more my nerves began to set in. I kept myself focused and my mind positive, as I prepared and then listened to music and relaxed with some breathing exercises for the last half hour before the interview. My mom's surgery was expected to be done around the same time I would be starting my interview which was unsettling, but I had to stay focused.

The interview could not have gone better. Other than a brief moment when I felt the realization of everything going on and I froze, thankfully I regained my focus and finished well. They were interviewing candidates for the remainder of the week, but they hoped to have a decision that Friday. I had a good feeling about it but didn't attach myself to the outcome, so I used my manifestation skills, set the intention, and let it be.

The next day I went to the hospital and thankfully, everything went well with my mom's surgery, and she was released that morning. I was so grateful she was okay and that her recovery went as well as it could have. I said my goodbyes to my mom and aunts and headed back to Milwaukee, for the five-hour drive home. On the drive home that Thursday afternoon, I received an offer from the recruiter for my life-changing job as the senior technology contract management professional position, contracting with Humana. I was so excited that I had to pull over. I was overcome with happiness and relief that I even shed tears. This was a game-changing, life-altering opportunity, and it couldn't have come at a better time.

I started my remote position with Humana in November of 2022. It has been the biggest challenge of my life, with an intense amount of rapid learning in an entirely new area, and many moments when I questioned if I could continue on. Despite the moments I doubted myself, I have enjoyed the challenges I face every day. I'm grateful to have a team of kind, helpful, and supportive women, especially after such a turbulent experience at Caterpillar.

The position with Humana really helped me take back my power, grow my financial security, plus it allowed me to make some big life changes. It turned my uncertainty of what I wanted to do and where I wanted to go next into certainty. In April, 2023, I relocated back to my hometown, into a brand new beautiful, spacious apartment that I absolutely love.

Since moving back, I feel a sense of peace, happiness, and security that I had been longing for. It is where my family and core friends are and, internally, I know it was the next right move for me. I was faced with uncertainty, challenges, betrayals, and devastation throughout my journey that ultimately led me back home. Not knowing what I wanted to do next—the unsettling experience at Caterpillar, my mother's health scares, the employment offer scam, and subsequent legal battle—left me feeling momentarily powerless. On top of all that, I felt crammed in my tiny, transition apartment in downtown Milwaukee which made it difficult for me to feel free and create.

Feeling my power stripped, lit a fire inside me to rise above it all and step into the most powerful version of myself. I learned to slay challenges, adversity, devastation, and toxic people, and use it as fuel to rocket me right where I wanted to be. This version of me was stronger, happier, more secure, and abundant than ever because I no longer accepted anything less than the best for myself. I was feeling the most powerful I had ever felt in my life. If it wasn't for the hurdles, intense challenges, and devastation I had to overcome, grow, and learn from, I would not have had the confidence to step into my best self.

The experience with Caterpillar, though intensely stressful, forced me to learn so much in a short period of time and pushed me out of my comfort zone every day. It gave me the confidence boost I needed, going into a much more complex and higher-level position.

Securing a life-changing role catapulted my relocation, peace, happiness, abundance, and security. Through the hurdles and painful experiences, I gained the strength, resilience, and confidence, to take back my power and step into my most powerful self. I am manifesting all of my dreams and accepting nothing less than the best for myself, in all areas of my life.

How did I take back my power?

I learned how to use upsets, setbacks, adversity, and pain as fuel to bulldoze through any obstacles that come in my path, while believing in myself and the direction my destiny takes me.

In loving memory of my beloved Coco, who was always there for me, filling my life with joy, love, and light through it all. You will forever remain in my heart.

About Natalie

Natalie was born in Kenosha, Wisconsin, and grew up in Eau Claire, Wisconsin, with her mother Linda and her twin brother Matt. She competed as a gymnast at the club level and for Memorial High School, where she earned an athletic award for Most Improved. Natalie also played tennis for Memorial High School, where she never lost a competitive match. She obtained her associate degree in Paralegal Studies in Eau Claire, before relocating to Milwaukee, Wisconsin.

After relocating to Milwaukee, Natalie obtained her Bachelor of Arts Degree in Business Administration. She was employed in various paralegal/legal roles, instructed gymnastics, and bartended at the Fiserv Forum arena, while living in Milwaukee. Natalie relocated to Tampa, Florida in 2016, where she relished the beach life and the thrill of starting fresh in a new city. After moving back to Wisconsin to be near family and friends, she transitioned into Information Technology after landing a position as a Senior Technology Contract Management Professional, in November 2022.

Natalie believes in the power of standing by one's truth and honor, being authentic, and always trying to do the right thing. She grew up from humble beginnings, with minimal parental guidance. Natalie used the odds stacked against her as a catalyst to prove to herself she could do anything, as long as she believed in herself and never gave up. Navigating through life's difficulties, she grasped how to plow through adversity and setbacks, both professionally and personally, with an unwavering resilience and warrior spirit. Natalie thrives on the challenge of pushing herself beyond her expectations, both in her personal and professional life. She learned how to use pain, grief, failure, betrayal, and regret as motivation to climb back up from any fall or setback and come back even stronger. Natalie enjoys helping others find hope, inspiration, and positive solutions while dealing with life's everyday challenges.

Natalie has a passion for writing, gymnastics, justice, astrology, learning and evolving, and discovering her life purpose. She loves spending time with family and friends, playing tennis, live music, running, attending sporting events, traveling, dancing, cooking, baking, reading, writing, and learning about psychology, astrology, and the law.

Natalie had the privilege of co-authoring an inspirational book in 2019, called *Manifesting Your Dreams*, which reached number #1 on Amazon. Writing is a creative and cathartic outlet for her, and she plans to continue to strengthen her skills and write her own book one day—hoping to inspire others.

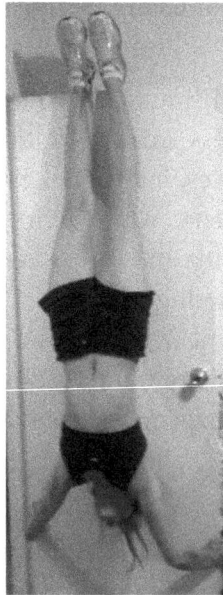

"Dream without fear. Love without limits." - Dilip Bathija

"We are only limited to the extent our minds dictate."
- Natalie M. Miller

"You gain strength, courage and confidence by every experience in which you really stop to look fear in the face. You must do the thing you think you cannot do."

- Eleanor Roosevelt

"In the middle of difficulty lies opportunity." - Albert Einstein

"The meaning of life is to find your gift. The purpose of life is to give it away."
- Pablo Picasso

STEFAN RYBAK

The phrase "Do not be afraid" is written in the Bible 365 times. That's a daily reminder from God to be fearless every day!

A Story of Strength, Courage, and Faith

One of the biggest lessons I've learned since my "surprise" unexpected heart operation is that what I don't know about my mother's life is often just as relevant and vital as what I do know. I have learned a lot from the facts I know for certain about her, but I have also learned a lot from the gaps. And sometimes, I have learned the most from things that she chose not to tell me when she was alive. It's as if my mother's long life was like a giant and complicated jigsaw puzzle, with a significant number of missing and broken pieces—very frustrating. I wondered, as I tried to put the puzzle together, whether some pieces were thrown away on purpose.

Perhaps all lives are like that, incomplete jigsaw puzzles left behind. We cannot expect the living to fill in all of the gaps. No matter how much I think about my mother's life, no matter how carefully I review the bits and pieces of near and distant memories, the picture that emerges is not clear or definitive. It has many empty spots where the viewer must fill in the details. The portrait that emerges is a mixture of saint and sinner, of a very human person, someone who made mistakes but who also found a way to grow stronger in adversity, someone who did not lose her faith in God, even in the face of hardship, death, and disaster. It's a picture that has become much more important to me since I had my own brushes with such realities and since I realized that my mother was a survivor of what would today be called Post-Traumatic Stress Syndrome. Despite its many defects, hers is a picture worth saving and cherishing.

My mother's information that shows up in these pages comes from my personal experiences with her and from words she said to my siblings and me as we were growing up. This information is not the whole woman, but it's all the pieces of the puzzle I could find. My mother, Maria Olszak, was born on August 13, 1925, and raised in southeastern Poland, near a town called Sporniak.

When Hitler invaded Poland on September 1, 1939, my mother had just turned 14 years old. She grew up in a farming family and was a good student who worked hard to achieve good grades. She liked being with people but enjoyed private moments of thought and reflection.

My mother was sitting alone in a field one day, doing her homework, when she heard a loud noise in the sky. When she looked up in the distance, she saw a fleet of airplanes approaching; then she saw what looked like silver balls falling from the planes. It was the German Air Force bombing the town. My mother gathered her books and raced toward home. We do not know what my mother said to her parents, or anyone else, about what she'd seen. We do not know what her family did when the town was bombed and invaded by the Germans.

There is a gap in her life story now of about a year and a half. The next puzzle piece is well into the war, sometime in the latter half of 1941. One day, at the age of sixteen, my mother was walking to school when a gang of men wearing swastikas intercepted her before she could get there. She was kidnapped, along with many other young people, by Nazi forces. They took all the Polish youth they could find. The German army wanted to put them all to work. My mother never discussed the details of her capture. She did not tell us who captured her or how she felt when it happened. My mother did tell my eldest sister, Barbara, that she was shipped by train to a labor camp somewhere in Germany. She was one of many thousands of Poles who slaved in unwilling support of the Nazi war effort.

This part of the story, the part about her being kidnapped on the way to school and sent to a labor camp, only became apparent to me during a discussion that I had with Barbara. I felt a strange mixture of emotions when I heard it. Pride in the idea that my mother could survive being sent to such a place and shock that she had been exposed to it. I even felt a twinge of resentment that she could keep something that important from me. Maybe, in her own way, my mother was trying to protect me from this horror; after all, I was her youngest of six children, and she always referred to me as her baby, even after I became an adult. In my mother's eyes, I suppose that I was more vulnerable and more deserving of protection from the most brutal facts.

The truth is that my mother didn't like talking much about the horrors that World War II brought into her life. I'm sharing the moments that my siblings and I pieced together, over decades, from various discussions about what happened to her during the war. My mother's answers to those questions were typically not very long or detailed. It was not something she wanted to focus on or think about.

My mother saw pregnant women getting kicked in the stomach by Nazi officers wearing heavy leather boots at the labor camp. She saw those women dragged away to rooms where more torture awaited them. She could hardly imagine what would then happen to them, and she did not want to. She saw the terrible abuse of women and children by Nazi officers. She saw children knocked to the ground and pulled around by their hair. When she was pressed for details about what else she had seen, she would grow quiet and act as if she'd already said too much. It was as though she had made a mistake answering the question in the first place.

When we lived in Connecticut in the early 1960s, my sister, Barbara, who was 12 or 13 years old at the time, saw our mother play host to a group of seven or eight women she called "The Rabbits." It was a strange name for these women and reflected a strange common bond with my mother. They were women who, as young students, had been part of the Polish underground resistance, intent on defying the brutal Nazi regime. Captured by the Gestapo, they were sent to Ravensbrück, the concentration camp known as "Hitler's Hell for Women." This is where my mother was also sent. They were called "The Rabbits" because the Third Reich used them as laboratory animals for medical experiments. The Nazis had used their limbs to simulate war wounds; they infected those wounds with aggressive bacteria, wood chips, and glass, trying to cause gangrene, which they would then attempt to treat. They also experimented with removing and damaging nerves, muscles, and even bones in the legs. "The Rabbits" limped; their legs lacked any muscle and had deep indentations from surgical removal of bone and tendons.

Some of the women who survived knew my mother. Barbara recalls that my mother would have these ladies over for dinner. They made a point of telling Barbara their stories in the hope that they would not be forgotten. Barbara does not remember all of the details. She does recall that "The Rabbits" were as forthcoming as my mother was circumspect. One of these women showed Barbara her bare leg. There was a horrible scar where someone had taken away the muscle. At the time of these discussions, I was an infant. I did not learn about what went on in Ravensbrück until many years later, after my mother's death.

I can understand the decision to withhold this painful information from a young child. But my mother and I connected as adults, too. She and I had decades to talk about this. Somehow, she never got around to it, though I asked my fair share of questions about what had happened during this period.

What led her to compartmentalize this corner of her life and keep me from ever being part of it? I will never know for sure. So far as Barbara and I know, my mother avoided being subjected to medical experiments. We do not know how she managed that. She may have been considered too young for such barbarity, but that seems unlikely. The Nazi doctors didn't seem to have been all that picky. Perhaps she worked very hard. She was young and healthy. Maybe there was a system under which hard workers were spared that treatment. There may be some other explanation.

In any event, we know my mother was surrounded by this mayhem, this inhumanity, and we know she worked long hours in the labor camp at the age of 16. We know she saw things no 16-year-old should have seen. We don't know what kind of work she did there. We know her youth was over too soon. We know it was stolen from her. We know her real job was to survive.

When pressed for specifics on what she had gone through during these years, my mother, as I recall, only responded with generalities. She told us that she always prayed to God that she would find a way to escape all the difficulties that she faced during the war. She said her family was religious, and they all prayed for survival. We do not know when my mother got out of the labor camp. Even though she never told us, Barbara and I feel sure, based on our knowledge of her character, that she decided to get out and took some kind of action.

Whatever means my mother took to get out of Ravensbrück, she did so because she was a strong-willed person. I had so many experiences of seeing her in action, pursuing something she wanted, that I could not imagine her passively accepting her fate in such a horrific environment. She must have taken some action, had some plan. But whenever I would ask her about such things, I'd get annoying, vague responses like "Stefan, that was such a long time ago," or "Who can remember?" According to my sister, our mother said she went out of her way to erase most of her memories about that time. She would push back when Barbara asked for details by saying things like, "I made up my mind to survive and did everything I could. Most of all, I prayed and prayed to God. That's all I remember."

I only came to realize, after her death, that my mother had actively chosen not to remember certain aspects of her life, perhaps as a coping skill to deal with the various forms of the war-related trauma that she encountered. Forgetting was a big part of how she survived. As it turned out, there was a lot to forget. Her whole life was, to a significant degree, measured out in trauma.

Barbara said our mother got through her time in the Nazi labor camp because she knew God would be by her side. She willed herself to survive, and her number one survival mechanism was constant prayer. My mother had a sincere belief that God had a purpose for her life, and that, somehow, God would allow her to survive to fulfill that purpose.

That is all very obscure, of course. We know for sure that there came a day when a German officer selected my mother to serve as his maid and personal assistant. Upon winning this job, she was allowed to leave the labor camp and live in a cramped attic room of this officer's home. Her new job was actually considered to be a plum assignment, and there must have been a great deal of competition for it. We have no idea how she first heard about this "escape" from the labor camp, what the selection process was for selecting such workers, or what my mother might have done at the time to take advantage of it. Perhaps it was an answered prayer. Her life could be measured out in those, too.

We figure that my mother must have been around 17 years old when the officer selected her to work at his home. She had probably spent about nine months in Ravensbrück. When the German officer (whose name we will never know, but whom my mother called simply "the general") first came in contact with her, he would have noticed several advantages she brought to the position he needed to fill. My mother was young, healthy, and active. In addition to Polish, my mother spoke German and English, having studied these languages in school. My mother was bold, diligent, and purposeful. The photographs of my mother during this period show that she was also strikingly beautiful. Barbara and I agreed that it was probably a combination of all of these qualities that got her the job.

My mother was able to walk out of the labor camp alive, with her limbs, faculties, and at least some portion of her dignity intact, which is something many people who entered Ravensbrück were not able to do. She told Barbara that the general was kind to her and did not take advantage of her in any sexual way.

Maybe the general liked the idea of a nice-looking young woman working around the house. I suppose it's as easy to imagine that reality for her, as a 17-year-old kidnapped victim, as it is to believe any other. The only obstacle to it, I suppose, is the challenge I have of picturing a Nazi general showing kindness to someone attractive, indebted to him for her survival, and utterly under his control.

But when you are putting together a jigsaw puzzle, you must find a way to leave the blank spaces blank, not fixate on them too much, and perhaps even show some gratitude for what you do not know and may never know. I choose to picture the general as someone with a shred of honor, Nazi or no Nazi, someone who decided to show kindness, to the woman who would one day become my mother, rather than exploit her. But it takes an effort of will for me to do that. I'm willing to make that effort.

The attic room my mother lived in at the general's house had a low ceiling. It was freezing in the winter, and my mother shivered in the cold darkness of those nights. She told Barbara that she often dreamt of her father coming to visit, warming her up, comforting her, and praying with her. She said that the dreams were sometimes so vivid and lifelike that she often had difficulty distinguishing them from reality. Perhaps her father, far away, back in Poland, really was trying to reach out to her during these agonizing nights. Maybe these visions or dreams provided some kind of connection between the two. We do know, as my mother grew up, they were very close.

I've thought of those attic visions of my mother's often. I have no idea whether people really can connect from such a distance. I don't think there can be any final answer to such a question.

Sometimes I have pictured myself as the one in the attic, the one shivering in an unfamiliar bed. I imagine my mother visiting to comfort me, to remind me of her presence and her love, to encourage me to keep going, to develop and always take back my own personal power, to be strong and courageous in the face of insurmountable obstacles. She would also encourage me to always be a person of faith, to keep praying, and trusting that the right path forward would eventually reveal itself for me.

About Stefan

Stefan Rybak is an energetic, enthusiastic, positively-motivated, and highly-experienced multimedia management professional with a 48-year track record of proven success and expertise in radio, television, print, and digital media.

Stefan's background and experience includes, but is not limited to, journalism, communications, broadcasting, advertising, marketing, sales, sales management, management consulting, strategic thinking, research, business development, understanding consumer behavior, new media, literary writing, leadership, mentoring, and dynamic professional speaking.

Stefan has written over 500 published articles and won the *Billboard Magazine* "Program/Operations Director of the Year" Award two consecutive years. Stefan is also a certified life coach.

Stefan is the author of the book, *The Shadow On My Heart – Faith, Family, Forgiveness*, which vaulted into Amazon's Top 10 Best Seller List on its first day of release.

Stefan was born and raised in Waterbury, Connecticut, and has resided in Michigan and Arizona. He currently lives in the greater New York City area with his family.

Mama and Stefan at her nursing home.

Stefan's favorite picture of Mama.

TRACY FELDMAN

"Be ready when the luck happens."

– Ina Garten
The Barefoot Contessa

Believe in Yourself

My writing journey started with pens, paper, a keyboard, and the best assistant a newbie writer could ask for. But that person wasn't a person at all, nor was it AI. She was the voice in my head who became my writing partner and editor. I named her Carrie. "Get Carried Away" was the phrase stamped in gold on my pineapple printed writing notebook. Her voice also sounded a lot like the fictional Carrie Bradshaw.

Journey is the perfect word to describe the process of writing one's story. When I read this book's title, I intuitively knew I would be taking back my power while writing about it. Carrie's editor voice helped me develop my **PI or Personal Intelligence**. She has been my **CB-GPT** (instead of ChatGPT), taking my thought prompts, fact checking memories, and helping me find my voice to tell my first-ever published story. The voice we found is personal and authentically me. It grew louder with each writing and editing session by pointing out how I have grown in my lifetime.

The title of this book is very **powerful**. Thinking about it, I feel we use **STRONG** or other words to describe someone more than we use powerful, so maybe we don't understand power? When I looked up

POWER on dictionary.com, I saw that it can be a noun (20 entries), verb used with an object (5), verb phrase (2), and an adjective (4). Each entry offers synonyms and examples, making **POWER** a very broad subject.

Some of my powerful experiences have been connected to **INFLUENCE**. Throughout this chapter, I will mention being influenced by people, events, or feelings. Sometimes I was so moved I took immediate action. Influencers aren't new. We've always had role models who inspired us to learn, grow, and find new paths.

As I begin my story, I want to share that I am celebrating a special Silver Anniversary—25 birthdays have passed since I made one of my most powerful decisions. I chose to fight for my life when faced with an unexplainable potentially life-ending illness. That experience made me who I am today—a woman who is thriving, growing, and learning so I can conquer new challenges and achieve new goals.

My story:

Always Take the Scenic Route

Confidently and consistently, I take the scenic route. Life is never boring; inspiration is everywhere. As a dreamer and a creator, I like challenges and learning. Not only do they fuel my soul, I level up and new opportunities are unlocked. The comfort zone is not where I excel. Like most scenic route travelers, my path has been filled with many plot twists and forks in the road. For me, the term scenic route is a metaphor for weathering life's storms, understanding they clear paths for something better and enjoying those new opportunities. When the plot twists happen, there are two choices: Option one is worrying and letting stress weigh you down like an anchor. (If you're an avoider-type, drop that anchor with it's friends: worry and fear to all go down together.) Option two is conquering fear and worry—rising, finding joy, and thriving.

I noticed I felt powerless when I allowed fear and worry to own more real estate in my head than they deserved. My education and the classes that interested me tell me I have always been attracted to analyzing things. Add bonus points for using a microscope and other magnified lenses for closer examination. Those simple facts of my life made it very easy and natural for me to become an over-thinker who was often guilty of letting the what-if-it-goes-wrong scenarios overshadow the everything can go right ones.

Eventually, those feelings became paralyzing energy suckers that reduced the quality of my life. You get back what you put out in the universe; for more positive results, you must put positive energy in the world. Stop wasting your precious energy on thoughts and activities that don't yield desired results. Now I am more productive and able to enjoy everything more. I adopted the good habit of setting myself up for success and not inviting the undesirable guests to my party. Enjoy your walk on my route!

My story began May 17, 1967 on Long Island, New York. From the start, the scenic route was destined to be part of my life. I lived in my first house in Massapequa exactly one day after leaving the hospital. On day two, my parents moved us into our new home 35 miles east in another county. In 1973, my family moved to South Florida. I liked learning a lot. Early on I sensed knowledge was power. School memories are filled with science and math classrooms. They were my favorite subjects. I had a photographic memory back then and did not have to study to achieve good grades. Most of school was not challenging for me, my mind was always thinking of something bigger than what I was being taught. My most powerful feelings I experienced were not fitting in and being an outsider throughout grades one to 12. I was on a quest for something where I would fit in and feel empowered. During school vacations, we visited family in NY. Vacations were filled with two influential F Words—**Fashion** and **Family**.

Fashion

NY fashions were different, always ahead of Florida … especially designer. When my Florida schoolmates started wearing designer jeans, I was ahead of the trend and wore brightly-colored Gloria Vanderbilt jeans purchased in New York. The colors were pretty and bright; the fabric was different than denim, plus it stretched. I took notice and was already being influenced by fashion. Those Gloria Vanderbilt jeans planted the seeds for my future career.

Family

My two older cousins inspired me. Robin loved animals and was interested in a veterinary career. Kelly was also a math/science fan who loved mechanical things. She saw that traditional choices were not enough, and successfully fought for more options for girls throughout her childhood.

I also observed my grandparents working. Their jobs were not like any of my friends' grandparents. My grandfather was a leading golf course architect/builder during Long Island's golf boom. He built a golf course in 90 days, and was a Navy salvage diver during WWII. His experiences, creativity, and work ethic have inspired me to challenge myself and work through anything.

My grandmother was a trailblazer as she ran the golf business from her home office starting in the late 1950s. I saw my grandmother working with her accountant, negotiating bank loans, and more. She was my first real-life role model with her confident, smart, and well-dressed business style. Family archives include copies of public relations letters from the 1960s she wrote to golf editors, her and my grandmother blended business name (JenMar equipment company.) At the time, I was too young to realize Jennie Martin had a successful career. I didn't know how important she was to the business' success. She was my wonderful grandma, and I love spending time with her … I am thankful my mom sent me to New York for this wonderful experience that normalized women achieving success in non-traditional ways. My mom must have realized early on how much I would benefit from spending time with her parents.

High school days were spent in Jupiter, Florida. I finally felt challenged by math, science, and French classes. My French Teacher, Glenda Wallin, inspired me the most. She was smart, confident, and brought a different vibe to learning than all the other teachers. Soon there were other women to inspire me when MTV (Music Television) debuted on August 1, 1981. Videos were creative, filled with fashion, art, and cultural trends. Strong female musicians became the **influencers** of my generation. They were inspiring me at a "contemporary" level.

For college, I was accepted into the fashion buying and merchandising program at the Fashion Institute of Technology (FIT), a State University of New York. I would be living and going to school in NYC. Adult life began the first week of September 1985 with my living my fashion buyer dream life, the one I planned for during high school. *Cue those clichés about the plans we make.* During my first hour of university, a big plot twist happened. Professor Ingrid Johnson was teaching Textile Science. Those first 50 minutes changed my life. I saw myself in her, and her intro into textiles mesmerized me. Math, science, and fashion collided as this dynamic, strong professor spoke. I immediately went to the registrar's office and changed my major to textiles technology.

At FIT, I had paid internships to help me gain work experience and find my career path. The most meaningful internship started during the summer of 1986. I was an assistant in the fabric buying department for Warnaco Women's Wear on 7th Avenue. My Warnaco boss Roberta Saft started her own business and hired me. After my third year of college, I was ready for my own studio apartment. I took a subway to Queens by myself, met with a broker, and signed a very affordable lease the same day.

Roberta later helped me get a job at Tex-Loom Industries with her friend Al Sandler. His first customer was his friend Liz Claiborne, who was starting her company too. Al was a wonderful boss, and I felt this would be a great long-term position. One December 1990 night after our company holiday party, I received a call from a co-worker. He said Al died at his home that night. All day Al had been spending time with each of us as he handed out Christmas bonuses. He was complimenting us, talking about growth and the future. We all lost an incredibly special person and mentor that night.

Tex-Loom was purchased by a businessperson with no textile experience who took a gamble. The business was never the same, and the company filed for Chapter 11 bankruptcy. I left, before the Chapter 7 filing, for a job at a company that would soon close too. It was a sign of the times and foreshadowed many downturns and changes that would follow throughout my career.

It was 1994, and my route was about to get even more scenic. I was now Mrs. Feldman, a new homeowner, and jobless in my shrinking dream industry with no prospects. Shortly before I left Tex-Loom, a large breast lump appeared. I had surgery. Thankfully, my lump was not cancerous, but it left a large scar, and a surprise medical bill. I made my cobra payments to Tex-Loom who cashed my checks, but the bankruptcy managers never paid the insurance premium.

The enormous bill was now my responsibility. I fought with no results. One day while looking for jobs in Women's Wear Daily, I saw TexLoom's **FINAL** bankruptcy notice. All claims were to be filed in Manhattan's bankruptcy court by a date that was two days away. I seized this opportunity. On the last day, my husband drove me to the courthouse to file my paperwork with only one hour to spare. I was now a creditor Tex-Loom owed with as much power as the large creditors. They could not settle the bankruptcy until my claim was settled.

My medical bill was settled in bankruptcy court, but I was still jobless. It was scary and frustrating to have things fall apart when I was doing everything right. That was my first time I recall feeling powerless. My husband took a lot of that feeling away because he had a good job and we were able to pay our bills. But he was an accountant, not a textile or fashion person. He had no friends or clients who had a job for me. It was up to me to make something happen and pivot long before pivoting was a thing.

I took classes in a CAD (computer aided design) at Nassau Community College to give myself more career opportunities. Months later, an ad for an assistant studio manager at Carna Mills was advertised. I wasn't sure what the job was, but it was the only one listed for months. Getting an interview was a priority even if it was not my ideal job.

It would be the first time where I thought it might be okay to settle for less than what I wanted. With limited prospects, it was landing that job or finding a new career. Luckily, the job fit my experience. However, I received a rejection phone call from my interviewer. He said, "I like you, BUT NO, you are so overqualified, you can do my job." After hearing that double-edged rejection, I did something else I had never done before. I confidently and **successfully** begged for a chance. The job was a great experience filled with lots of opportunities.

February 1999 brought me a bad lingering flu for Valentine's Day along with a life-changing health crisis. One day riding the Long Island Railroad home, 20 minutes into a 50-minute express ride, my heart started beating quickly. I felt like I was having a heart attack. The pain was crushing, and the heartbeats were intense and erratic. I thought of my husband, our life and **fought hard** to calm myself so I would make it to my train stop. Then I would instruct my mom on how to drive me to the cardiac hospital (she didn't know the route.) I did not want the train to make an emergency stop and go to a random hospital with no way to contact my husband or mom.

Friends, please read the paragraph above again as if it were you on that train and your heart beats were louder than all the train noises. **Read** it again only stronger this time. Do you feel the urgency? Let me add more tension by telling you that the ER (at the best Long Island heart hospital at the time) I was planing on driving to was another 30 to 40 minutes away. While struggling to calm myself, my overthinking skills came in really handy with my survival plan. I distracted myself between stations. As we passed Lynbrook, Rockville Centre, Baldwin, Freeport, and Merrick, I was that much closer to getting off in Bellmore. At the same time, I was planning for alternate routes in case of accidents on the roads. Instead of fear, my thoughts were *I am young with no heart problems, how can I be having a heart attack?* I was determined that my life was not going to end on that train or alone in a strange hospital.

There were no traffic issues on the way to the St. Francis emergency room. I was seen right away, stabilized, and got the basic tests which did not show anything. I would need an electrophysiology test the next day. St. Francis was overcrowded so I would spend the night in a large room with eight other cardiac patients who were 30 to 50+ years older than me. I remember thinking I do not belong here as I heard the conversations among my roommates and their children who were probably older than me too.

Cringe warning!! During my EP test, I was fully awake and could feel the device travel across my body into my heart. Adrenaline was added to my IV while the exam table was slowly being flipped upside down. My cardiologist was trying to replicate the same heart racing experience without my passing out. I needed to be alert so I could say I was about to pass out. At that point, the procedure was stopped. The doctor did not find an electrical issue so there was nothing to fix. His discharge instructions were to be careful, avoid caffeine and any stimulants that would cause a rapid heartbeat, and seek help if you experience this again. "We have no answers" was not a comforting statement.

The experience fried my already stressed nerves. I am positive that test did more harm than good as it triggered an endless medical battle. My body was still recovering from the flu, and I had an adrenaline infusion, and more trauma added. I kept having the racing heart and was getting sicker. The heart racing made it feel like I was going to die. My body was attacking me and breaking down. Each week brought a new health crisis, more doctors, more specialists, more tests, and no answers. Every system in my body was failing and being tested. I kept getting no answers, just more medication that was making me worse, not better. I was down to 100-110 pounds which was life-threatening at 5'9". I was not anorexic, but looked like I was.

A few weeks later, the racing heart issues did have a notable change as my body went into panic mode. I was so unhealthy that I started to fear sleep. I felt I would die when I went to sleep. Fear of dying led to a feeling of powerlessness and severe insomnia. At age 31, my scenic route was looking like a dead end (pun intended) or a lifetime of medical problems. I was always so energetic and healthy. Throughout my illness, I tried to be calm and brave. I did not let my husband, Greg, know how sick I was; he always saw me as strong and confident, not frail.

Patients MUST focus on healing and surviving. Spouses can become lost and powerless in the process. It is not intentional, but this new "CRISIS" reality changes the relationship dynamics. Twenty-five years ago, the support resources for spouses and family were smaller or non-existent. Today there is an abundance of support, and I encourage you to utilize those resources if you are experiencing any crisis.

My Carna Mills boss, Jordan Rose now witnessed me deteriorating for months and suggested I visit his longtime doctor friend for help. He warned me that Dr. Ronald Bland was an alternative healer and to have faith he would help me. Jordan's exact words were "Ronnie finds answers no one else can." I had nothing to lose. My choices were death or paralyzing bad health while using Zoloft and other sedatives to live a life that was nothing like the one a vibrant, healthy person would live.

I left my first Dr. B. appointment with some symptoms relieved and a kinesiology/chiropractic treatment plan to restore my health. That plan also included a severely restricted anti-inflammatory diet so my body could heal without having to waste energy processing the wrong foods. I was finally on the road to recovery. The insomnia was not so easy to resolve. For almost a year, I existed by sleeping 10-15 hours per week. I tried to resume a normal life, but sleep deprivation made it hard. Calming my mind was something I would have to master, and it would take time.

Eventually, I got my sleep under control, and I now make sure to never do anything to jeopardize that recovery. Sleep and good health are non-negotiables. I learned so much about my mind and body during my severe illness. The power of positive thinking and taking control of your own health is enormous. I pay attention to the warning signs of stress and unhealthy habits. My healthful routine now includes what Dr. B. taught me plus additional practices I learned from others who had their own health battles.

The best answer I have for why I had a heart problem is the combination of a bad flu strain and using over-the-counter decongestant cold/flu medicine as a mitral valve prolapse patient. Doctors previously discovered I had mitral valve prolapse during a routine physical and never told me not to take decongestants. As an MVP patient, I was routinely prescribed antibiotics prior to dental appointments to prevent endocarditis but did not receive other warnings. That antibiotic practice was discontinued in 2007 because it was found to be ineffective. I remember so much of my ER visit very clearly, getting blood work is not one of the memories. If my blood was tested, the troponin test was newish, there were reports of false results, or it may not have been done. I have read conflicting dates for when the test was put into use, they are around the time I got sick.

Once my health was restored, I was able to focus at work on the new German textile software called Tex-Design we were purchasing. I asked to learn this software with my design team. Soon all but one designer left, and I was now designing plus doing my own job. It was exhausting, but I liked it. The business was declining and changing in ways that the owners could not keep up with. I never thought to say "no" to this additional responsibility. My industry experience so far told me I should do more with computerized designs.

Flash forward to September 11, 2001. I was looking out my 6th Avenue office windows to the sun shining on the Empire State Building when WPLJ disc jockeys Scott Shannon and Todd Petengill announced that a plane flew into the World Trade Center. With Manhattan burning behind me, I walked uptown to the Park Avenue Armory to meet my husband who was activated with the New York State Guard at that location. Approximately two months after 9/11, Carna Mills decided to phase out their business, and I was let go. I did request to get one of the design computers and the Tex-Design software when I left. The business would not need them anymore, and I figured I could use them to design and freelance while trying to get a new job. Tex-Design's New York distributor hired me to train their customers' designers. I got to travel, learn additional modules in the program my company did not use, and teach others.

Beside work, I have put my creative and technical skills to use in other ways. One was volunteering for a crochet project at the Long Island Museum. Fiber artist Carol Hummel was seeking community volunteers for her museum installation. My mom and my grandmother crocheted. But I never attempted it until that project. Carol is known as the mother of yarn bombing. She had the same impact on me as Professor Ingrid Johnson did at FIT. Both ladies knew their craft and were smart, kind, and eager to share their talents.

A year later, a different local museum asked for yarn bomb volunteers. Museum Director Lauren Hubbard admired Carol's trees and wanted to do something for her Maker's Faire and Maritime Explorium Museum. Eight crocheters attended the first volunteer meeting thinking we would repeat the same process, pick up yarn, instructions, and then return a crocheted piece. None of that happened. Lauren did not know how the trees project worked or that the artist was from Ohio. She mentioned a few ideas, and it was clear the volunteer group would need to create everything.

Soon, I said "I am a textile designer, and can create a mockup for our next meeting to see where we go." I went from crocheting a circle using someone else's instruction to creating a design and organizing an art installation for a children's museum in less than an hour. This design was going to be so different from Carol's art. My creation came to life using the museum's logo colors, a six-foot octopus, and six-foot seahorse. The two yarn bomb sea creatures became mascots named Portia and Jeffrey in honor of the museum's hometown of Port Jefferson. My mom was Portia's crocheter. There were many engineering challenges with the design and installing it on a historic museum building. The team worked everything out as my vision came to life. I was given an extra assignment, teaching an interactive yarn bomb class for the Maker Faire. The class was a success, and this installation brought joy to museum visitors and the public for over two years.

Lessons Learned, Footnotes, and Conclusions:

- Conquer fear and worry. My husband Greg always said, "What does it matter?" when I started to overthink. He taught me I was giving away my power when I worried about things I could not control, and I was putting energy into things which added no value to my/our life. Greg's favorite phrase is a prompt to **PAUSE**, and remember life will go on everyday regardless of how much I stress. Focus on what I can control (my attitude), and have fun.

- Harness your superpowers. All the years I felt different and didn't fit, I was really me discovering mine. We each have our own superpowers—the world needs us to use those talents.

- Only you can take back your power. No one can do that work for you. This is your life, so make it the best life for you. Find people who inspire you to level up.

- Get LOUD ... find your voice to stand up for your health, happiness, and life goals. Knowledge and truth are essential when navigating disruptive plot twists.

- Eliminate negativity and toxic thoughts—they pollute your positive energy flow. No negative or limiting words about yourself. Do not let negative thoughts start your day or tasks. You are placing one more obstacle in your way as a successful self-power manager.

- Once you restore your power, don't give it away.

- Positive attracts positive and repels negative.

- Pay attention to repeated themes in your life. They are important clues.

- My daily go to mantras:

  o Believe in Yourself: you have survived death and can do hard things. My dad taught me this lesson early when he led me to highway I95 just 20 minutes into my first driving lesson.

  o Shake Up Your Snow Globe: change is created by shaking things up. No limiting ideas allowed.

  o Today is Someday: go try new things, take that fork in the scenic route, and you will find something better than you imagined.

There are updates on my health crisis. I now believe I had a rare mild heart attack. In 2024, I had met and also heard of young women who had mild heart attacks with no personal or family heart disease history, no drug use and good health. Their symptoms were the same as mine, and they had to fight for treatment too. The thought is healthy young women don't have heart attacks, they have panic attacks. More tests are available today, but you still have to self-advocate if you are not in the target group.

Not being diagnosed with a heart attack was a blessing I am extremely thankful for. My whole life would be different if I was put on heart medication and never learned alternative methods. Lots of important characters would be missing from my story if I took a different path. The experience helped me create effective health plans for myself and my dog when traditional methods weren't working. I even use that knowledge in caring for my elderly parents today. We all use vitamin IV therapy, acupuncture, and Chinese medicine with success.

During the October 2024 Unwrap Your Best Self women's wellness event in Cocoa, Florida, I told my story on a speaker panel at the event. Afterward, I asked one of the presenting doctors (Dr. J. Ann Dunn, a kinesiologist) to sign her book for me. As we spoke, Dr. J. asked me who my New York doctor was. When I said Dr. Ronnie Bland, she smiled and told me he was her student. Talk about something being kismet or bashert!!! My crisis still guides me toward the path I am supposed to be on.

This chapter is a lifetime in the making with several pauses and edits. During the first draft of editing phase, I did a lot of living on my scenic route and learned from many on my travels. I also created more art and took classes again. One notable art teacher, Midge Baudouin reminded me to "Pause, take a step back from the canvas, and it will be even better." Midge calls me the textile scientist at her classes. My photographer friend Olivia Womack taught me so much in 2024 with my EMPOWERING portrait session. Olivia captured the power of my life so far, and highlighted a future filled with possibilities. She invited me to speak at her Unwrap Your Best Self panel. The public speaking experience was a first for both of us, and it was amazing to tell our stories out loud.

Friends and support are vital to maintaining your power balance. I learned the importance of help and community. Special thanks to Stacy Kaplan, Carolyn Blackmon, Holly Shelley, Ginger Chouinard, and Anna Feldman for all the long phone calls. Greg, you proposed as Santa saying the last gift in the Santa sack was going to be you. What a gift you have been! You have taught me more than anyone in my life, given me opportunities I never knew I needed, helped break bad habits, made me stronger, and you amaze me with your brilliance and generosity.

My "Friends in Overdrive" list is long and growing. The original FIOS Theresa, Claudia, Mary, Kat, Karen, Aimee, Sara, Donna have given me smile lines for years. Eva, Bobbi, Olivia, Christine, Rebecca, Kenny Klein and others have been there for the newer chapters. Robert always has the kindest words for birthdays. Robin, Kelly, my parents and the rest of my large family have been supporters for the long haul.

About Tracy

Tracy was born in Huntington, New York and grew up in South Florida. After high school, Tracy returned to New York where she earned her Associate of Science degree in Textile Science and Technology; and a Bachelor of Science degree in Marketing, Textiles Specialty.

As a two-time Cotton Inc. award winner, Tracy has created textiles for apparel and home furnishings in India, the U.S., and many countries throughout the world. You have already read about some of the companies she has worked for during her career and how she embraced new opportunities. By pivoting over the years, Tracy has added skills in sales, marketing, promotion, order and inventory management, product development, packaging design, product photography, photo editing, catalog, and line sheet design, website management and branding to her wheelhouse. She believes learning is essential for being ready when the luck happens.

Because of her health crisis, Tracy advocates healthy practices, stress management, clean eating, and limiting toxins and chemicals from her life where possible. She uses a combination of traditional medicine, alternative methods, IV vitamin therapy, and physical therapy/exercise. Research, maintaining good records, recommendations and proceeding cautiously are all part of her approach as a caregiver. With planning, Tracy has found doctors for her parents who are conservative in their approach to prescription medicines and believe alternative methods can be used to maintain quality of life as we age. Please consult your own doctors, proceed cautiously, and do your research for the best plan for you; make adjustments as needed. It is essential to make sure elements on your plan are compatible and don't negate each other.

Following in her grandmother's pumps, Tracy is now involved in her family's golf business at Spring Lake and Swan Lake's Golf Courses on Long Island. As Swan Lake's VP, Tracy has been a virtual golf lobbyist for National Golf Day, hosted Women's Golf Day events, and belongs to several women in golf business groups. In June of 2024, Tracy became the Space Cost Market Leader for Women on Course, a golf networking group redefining how modern women play golf.

Travel, art, design, museums, theater, music, reading, walking, 5Ks, yoga, golf, rescuing dogs, animals, marine life, philanthropy, and women's leadership/empowerment are all things Tracy enjoys. She currently lives on Florida's Space Coast and is inspired by Florida's FIT and their WE Venture women's business school. 2024 was her third year participating in WE Venture's Wine, Women & Shoes fundraising event.

Power rules Tracy lives by:

- Believe in Yourself!
- Dream Bigger than your current ability, and do the impossible.
- Never stop learning; try new things.
- Tell your story, it will change someone's life.
- Praise and inspire your team.

There are many things Tracy hasn't done. Stay tuned as she crosses them off the list! When life gets tough, use her light, and have faith that you'll be ready when the luck happens!

Follow Tracy's blog, *The Lovely Spotted Flamingo*. She is thankful for her family, friends, rescue dogs, flamingos, turtles, dolphins, 80's music, "The Marvelous Mrs. Maisel," and this *Take Back Your Power* opportunity!

Tracy Feldman Co.
swanlakegolf.com
springlakegolfclub.com
@almost_heaven_emerald_isle_nc on the 'gram
thelovelyspottedflamingo.com
P & J Yarn Bombers Facebook Page
powherfulliving.com womenoncourse.com
twolumpsofsugar.net

Portia and Jeffrey Yarn Bomb installation, Maritime Explorium Museum, Port Jefferson, New York.

Spring 1999 as Tracy's health was deteriorating quickly.

Lemur yoga at Big Cat Habitat, Sarasota, Florida.

Tracy with her cousins Kelly Roy and Robin Parker.

Believe in Yourself portrait.
Photo credit: Olivia Womack Photography

EL PELLEGRINO

"Life is a daring adventure,
or it is nothing at all."

– Helen Keller

My Pathway to
"The Age of New Beginnings"
Transformation From Darkness to Light

I am blessed and honored to describe my spiritual journey from fear and darkness to love and light.

I am known as Priestess El. I have been a lightworker and spiritual counselor for over 30 years. I help people let go of past traumas so they can transform, heal, and live their best life. We all need to let go of what does not serve us so that we can become our highest and best selves. I am committed to healing and spiritual enlightenment because as a child I suffered trauma which left me full of anxiety and fear. I transformed my limiting emotions to love, light, and hope creating endless possibilities in my life and the lives of others.

When I was a little girl, I had very brilliant and eccentric parents. Like many people, my childhood was not perfect. My parents had me in their later 40s, and although both of my parents were successful, they did not create a healthy environment for me.

It was my mom's third marriage and my dad's first. They had a tumultuous relationship that negatively affected me in my most formative years. Both of my parents made it clear to me that children were meant to be seen and not heard. Even when I was in my crib, they would ignore my screams and let me cry myself to sleep. I learned from an early age that my needs and wants didn't matter at all. They were both completely unaware that their child needed to be comforted to form a healthy sense of herself in this world. In all fairness to my parents, this behavior was the norm for most adults parenting children in those days. Now, it is known that from birth to five years old, children become hardwired in their emotional patterns, their perception of themselves, and how they perceive others.

Unfortunately, both of my parents had their own demons; there was constant turmoil, violence, mental, and physical abuse. I remember huddling in the bathroom, sitting on the cold tile, pressed against the door as my parents fought in the other room.

My father was drunk, smashing my mother's beautiful artwork, as my mother screamed in rage. Terror filled my small, shivering frame. It was inevitable that their anger would eventually be directed toward me. I was alone with no one to protect me. It was like being on a roller coaster ride every day. There were ups and downs, and I never knew what was going to happen next.

It was a common occurrence to have the police show up at our door due to the fighting. When the police arrived and the drama came to a halt, there was still an unsettling feeling of adrenaline coursing through my veins as I tried to unwind and settle back down into a strange superficial calm. I would compare this feeling to how one would feel after almost experiencing a terrible car accident shaken and bewildered. I wanted to be invisible, to fade away, to die, because the pain I felt inside of me was too great to bear.

My mom would go from zero to 60 without warning—at just the drop of a hat. There were times I would lay in bed with her, after she fought with my dad, afraid to even move a muscle. I can remember lying there listening to the sounds of my heartbeat afraid that if I moved, even just a little bit, it would set her off. She would yell at the top of her lungs, "What are you trying to do, torture me?" She was brilliant but had some serious mental health issues that forced me to always feel uncertain and unsafe, like I was walking on eggshells.

As I got older, tensions eased up a bit. When I was only a mere 12 years old, my mom treated me as a friend and brought me to dance clubs with her. Now that I look back, I realize I had no right being in a night club. But it was fun and fulfilling to be acknowledged, included, and get all dolled up. I was thrown into a whole new exciting world. I would dress to impress and feel beautiful! There was no stopping me! I finally felt accepted. It was at this point that I realized it was time for me to become an adult and take hold of my future.

My parents were still fighting daily. Although they were wonderful in their own ways, they were toxic together. When my parents finally divorced, I knew that for me to survive, I had to leave my mom and that lifestyle. So, I decided to choose my father. We traveled to San Francisco and moved in with my Aunt Lila in Palo Alto.

I started school in California and began working in a health club across from Stanford University. I was able to get a California ID that said I was 21. This enabled me to get my bartending license, and I started bartending at parties working for high-end companies such as Microsoft and Macintosh. Both companies had just started up around that same time. I made friends quickly and found myself at a party with my newfound friends who unbeknownst to me were into really heavy drugs. I left the party with a young man so we could pick up another friend, and I didn't realize he was on drugs until he was peaking on mescaline and started driving erratically. The car flipped six times, and I flew out of the back window. The accident left me in a coma.

After about a week, I woke up from my coma and never felt so alone. My father and aunt were so used to my independent behavior that they had no idea I was even in the hospital. They didn't worry at all about me in my absence. I woke up in physical pain, and no one was there for me. My whole world was shattered.

Did no one look for me? What world was I living in?

My dad left at this point to go visit family in Arizona and didn't even think twice about not saying goodbye to me. Meanwhile, the doctors had to pump my stomach from internal bleeding, and I could barely move with six broken vertebrae in my back. When I came to, I contacted my aunt and cousins who wanted nothing to do with me, as they were proper, and my behavior was appalling to them.

By age 15, I had already worked so much that the hospital was able to set me up with disability and unemployment. I was able to find an apartment on the boardwalk of Mission Beach, San Diego.

After all of that chaos, a miracle happened. I went back to New York to visit my sister and my mom, and I met Anthony, a man who saved me from my childhood. He was not the wild type I was usually attracted to. His sincere, loving attention and generosity won my heart and soul at the young age of 16. He was so kind and sweet. Anything I asked for was mine. If he didn't have it, he would find a way to give, get, or create it for me. I felt safe and unconditional love for the first time in my life. The beginning of our relationship was like paradise on earth. I enjoyed every minute of our love-drunk romance. When he proposed to me, it was a definite "yes!" Finally, I was able to push down the negative feelings from my childhood and felt pure bliss. That man, my prince, is still my loving husband today, 41 years later. We have three beautiful children who have children of their own now.

Though my husband and I have an unbreakable bond, it wasn't all a fairy tale. My husband came from a very loving family with a controlling Italian mother. She didn't want to give or share her son with a young, vivacious Jewish blonde. I think most people have seen the television comedy "Everyone Loves Raymond," and can remember how Ray's mom treated Deb. Well, imagine Ray's mom, Marie Barone, on steroids! I wasn't good enough for her son or welcomed as a family member. Anthony's mother made it very clear that I was inept. I could never measure up to her standards. It wasn't until the last few years of her life that she showed me any kindness or gratitude. Even though I suffered years of emotional abuse from her, when she fell ill, I took care of all her needs. She confessed to me then that I was an angel in her life. Not only because I was caring for her so lovingly, but also, when looking back, sharing my family with her was a blessing she truly valued. She was so grateful for her grandchildren, sharing the holidays, and allowing her to teach us the "Italian way."

Although hearing this from her was very healing, it didn't take back the years of feeling inadequate in her company. I was often the topic of family jokes; my role was undervalued and my character criticized. That emotional abuse and undeserving treatment lingered inside of me.

Her son, my husband, was not intentionally mean to me, but he was more than happy to put me in the role of "Lucy Ricardo," a silly housewife who was not taken very seriously. This made me feel like I was unworthy of true love and respect. I had a diminished sense of self, and I had started losing my self-confidence and self-worth. I was saying "sorry" all the time to my husband, just like I used to say to my mom. In a desperate attempt to hear I was worthy and loved, I would ask my husband constantly if he loved me. Every day, I was sadly grasping at straws to feel loved and wanted. I yearned for our "honeymoon phase" back.

I was also dedicated to our family businesses which eventually became catering halls. I helped design the rooms, the window treatments, and the lavish decor. I went above and beyond to fill every role that was needed from waitressing to running parties and bartending. My contributions to the businesses always seemed to go unnoticed as I worked without pay or any credit.

Since early childhood, there was always a part of me that just wanted to be invisible, to fade away, and to leave this earth because the pain inside of me was too great. I started to feel like that little girl again and wanted to escape from my feelings of humiliation, rejection, and despair. I pushed on and repressed my feelings, as I put all my energy into the business and my family. My inner turmoil manifested into several years of deadly and severe life-threatening physical accidents.

Soon, after giving birth to my second daughter, I fell, hit, and fractured my tailbone so hard that it caused a nerve problem called coccydynia. Coccydynia is a hyper-sensitive nerve disorder that results in whole body pain. From that point on, years of these freak accidents continued. A limo driver, not realizing that I was looking into the trunk, slammed the trunk down on my head causing skeletal damage, compressing my skull and neck down three vertebrae. This caused more nerve pain along with memory and hearing loss. A year later, I fell on ice and smacked the back of my head once again. The following year, I fell out of a moving catering van smashing my head on the highway. I was suffering a lot of physical pain, but I didn't want to upset anyone or ruin my family time, so once again I swallowed my feelings and tried to remain as quiet as I could about my pain and suffering.

One day I spoke to a close friend about my physical and spiritual pain, she lovingly listened to all I was going through. She felt my pain and despair and told me about a place that could help heal my heart. Since that day, I went on my journey of healing; I spent over 10 years in holistic healing courses such as Hypnotherapy, Intuitive Consciousness, Master An-ra, Ayurvedic Healing, Medical Mediumship, Emotional Freedom Technique, and so much more. I remember the day I began my journey like it was yesterday. These modalities provided me with the tools to begin my transformational journey and to heal the hurts of my past and present. I realized that I had to make changes for me to be free from the terror I felt in my heart. I needed to dig in deep and learn everything I could about how to heal the wounded parts of myself.

I sensed that if I accomplished a deep and profound healing, I could bring about greater love and light to the world. I now had a road map to heal myself and others. I studied everything I could on how to transform feelings of neglect into feelings of safety and contentment, self-love, and acceptance. I no longer needed to search for outside validation. I looked inward and upward for peace and stability. I no longer got into freak accidents or experienced regular physical trauma. I have healed my relationships and learned how to help hundreds and thousands of others do the same.

Since then, I magnified my power and purpose to another level. Letting go of fear and past hurts is the only way to channel love and light. Healing is a forever journey that may take twists and turns but is always evolving. Today I am at the point in my life where I can do quantum healing and DNA shifting without even a thought. The process involves channeled hypnotherapy along with intuitive work that helps people see life from a different perspective. I have held many classes, retreats, workshops, and seminars (locally and globally), which have helped others connect to their higher self, raise their vibrations, and find happiness.

I know now that my purpose in life is to help others through my system of healing. I help people shift out of negativity, and into higher frequencies of being so they can find freedom too. I help people reignite their desire to live and adopt the mantra: I now choose to live. I don't want to just exist. I choose to feel alive and live an extraordinary life!

Here is a testimony from a client who ascended to her highest self, "El gave me the strength, the confidence, and the loyalty of not giving up on me. She shifted all those negative fears of being alone, not being loved, and the fear of not having enough money to make ends meet. She introduced me to fabulous people along the way. I am laughing again the way I did as a child. True belly laughs! Due to her persistence, her loyalty, and her advice she made me a better woman; I changed the way I think, feel, and look. I will conquer the world, no doubt!"

If you would like to heal and are ready to feel deeper freedom, the first skill you need to awaken is to trust your intuition. Intuition is the divine connection to all that is. Every human being can connect to intuition as it is an inner knowing that flows through and to us all. We receive intuitive messages every day but most of us are unaware of this process. Intuition is learning to step aside and turn off your ego, conscience, and subconscious mind so you can receive the information that God and the universe are sending.

Children are very receptive to this type of information because their hearts and souls are open. They have not yet learned to judge, censor, or limit their beliefs to only what they can see or feel.

Intuition comes from direct communication with our higher self, our soul essence, God, and our universal consciousness. It is considered our sixth sense perceiving things without relying on touch, taste, smell, sight, or sound. Many identify the third eye as our intuitive center that works in concert with our pineal gland to help us connect to God and all that is. The gift of making your intuition an effective tool is learning how to bring unconscious data to a place in your conscious mind without judgment or filters. Just as emotions can shroud your logical thinking so can logic, reason, and feelings cloud your intuition. Intuition does not reason, it simply knows. It is perception without expectation.

How do you know when you are tapping into this eternal gift of sight? When you trust what you feel, not what you think. You are switching from intuition to reason when you start to think of correlations or use words like "should have, you know, or you could." Intuition is usually fragmented metaphors or pictures that tell you a story instantly; they do not follow a pattern. If you begin to think "if this happened, then this should happen" you once again are using reason. The fragmented pictures of intuition usually don't tell the full story alone, but when put together they are messages meant to be understood not pondered. It's important that you trust yourself. What you think, say, and do vibrates outward, branching indefinitely, thus formulating your reality which brings us to our next tool, the law of attraction.

Utilizing the law of attraction basically means what we say and think to the universe manifests, creates, and defines our reality. I help people find their highest and best thoughts so they can reach their highest potential. Anything that you think or feel on a regular basis will become reality, so we must have thoughts based in love and light, not fear and hate.

Love and acceptance will bring us to peace and abundance. Everything I am, is everything I'm supposed to be.

About El

El (Ellen) is an entrepreneur who has been happily married for 40 years. She is a mother of five and blessed with four wonderful grandchildren. Ellen and her husband, Anthony Pellegrino own Windows on The Lake and the Beach Club Estate, two prestigious catering halls in Lake Ronkonkoma, New York. She has been an event planner for over 35 years in their catering business. This has particularly been a special place in her heart where she brings people together to enjoy life, special occasions, and memorable moments of love.

Her goal is to continue to be an essential part of empowering people in the world. Over the past 30 years, Ellen has been working toward achieving this mission through her health and wellness business, the Age of New Beginnings. Ellen's business has helped others truly uncover their unique identity through soul searching and various coaching methods aimed to tap into one's true self and reveal a sense of purpose, healing, and ultimately, happiness.

The Age of New Beginnings signifies boldness, empowerment, and awareness to overcome challenges and explore "The New You" through transformation for inner peace and outer beauty, all while receiving hair styles, make-up treatments, massage, and facials that complement your lifestyle transition from the inside.

Ellen's true passion is giving back to the community and connecting through love and compassion. Her humanitarian work has caught the attention of Ronkonkoma Rotary International, in which she was honored with the Paul Harris Fellowship Award. Ellen was also acknowledged in her local community and honored with the People Who Make a Difference Award. Ellen and her husband, Anthony Pellegrino continue to support their community by donating annually to the Lt. Michael P. Murphy Run Around the Lake Ronkonkoma. She helped create the Historical Festival for Lake Ronkonkoma which celebrates the Annual Indian Princess of the Lake Festival occurring every June. This event showcases organic food, homemade goods, and holistic services.

Ellen produces and hosts the show "Age of New Beginnings." It challenges a person to take a chance on a new beginning of change, embracing light, spirituality from within, and channeling connections of other like-minded spirits and souls in need to reach their destiny.

Ellen enjoys helping and empowering others to transform to "A New Beginning" and new way of life, discovering their uniqueness and individuality. Her goal is to change one life one day at a time.

Through intuitive energetic healing, hypnotherapy sessions, well-being classes, and retreats, Ellen hosts events that bring people together to foster individual and community healing. By increasing awareness of others, to live life to their fullest, with a focus on moving forward through the law of attraction and higher universal consciousness. "Learning to live at a higher vibration removes blocks, obstacles, and allows one to be more in-tuned with their environment," Ellen says.

Ellen spent a few years with The UN during 2016-18 with Vigil for Peace. She participated in April-Earth Day 2024 in Times Square, NY and was on stage doing a multi-cultural event which aired in over 200 countries all around the world.

During the first month of the COVID-19 quarantine, she co-created a 22 Day of Transformation to help people cope with this devastating turn of events, which started on Earth Day. Ellen was part of many global zooms doing healing events during COVID-19. As it ended, she coordinated The New Life Expo on Long Island, NY—a wonderful holistic expo known for New Age Awareness and Wellness.

In the past six years, she has executive produced and completed four movies: *An Eddie Rocky Riveria Movie*, now appearing on Amazon Prime and $treetz 2; a crime drama, *The Next Generation of the Spanos* with Angel Salazar (Scarface, Carlito's Way) … appearing on both $Treetz; as well as *The Devil's Order*, A Star Hanson Film—a horror and comedy; and she is now finishing her next film, *The Good Samaritan*. Ellen utilizes all her staging, make-up, editing, and coordinating skills to produce an independent feature film for mass release.

Lastly, on March 22, 2023, El was presented with The Premier Business Women of Long Island Gala Award by Herald Newspapers at The Heritage Club Reception Hall in Old Bethpage for recognition of her accomplishments.

Ellen is referred to as Priestess El Pellegrino because she marries people at her catering facilities and provides healing work that elevates them to see the world from a higher prospective.

Today, she is honored to be part of this wonderful book, *Take Your Power Back*, with Marla McKenna and 20 other amazing people.

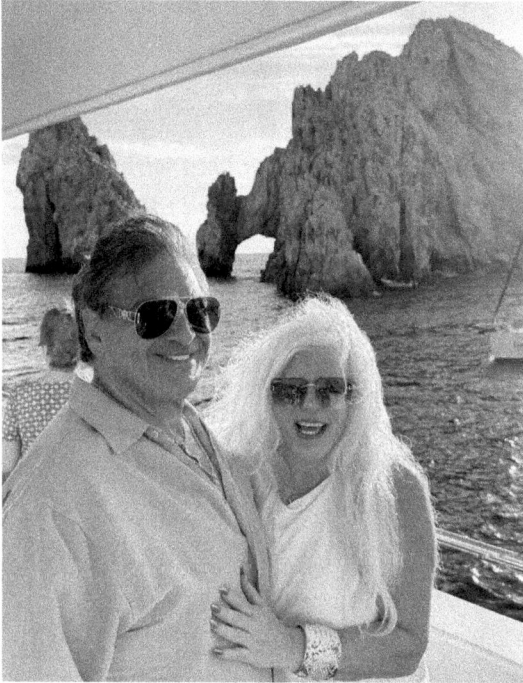

El with her husband, on their 40th anniversary getaway.

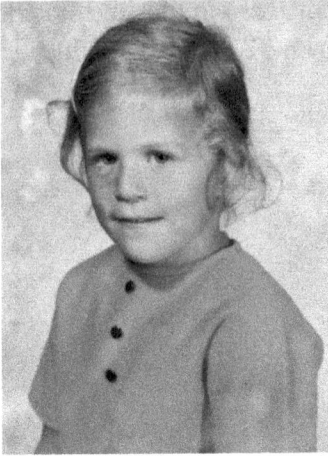

El, age 6, in kindergarten.

*El with her kids
and mother-in-law.*

El's kids in the 1990s.

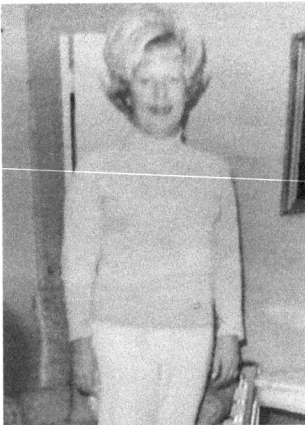

El's mom.

El and her dad.

*Christmas, 2024. El with two daughters,
son (with beard), and two of her five grandchildren.*

*El's grandchildren Raphael and
Mateo, who live in France.*

KAREN CARR

"You're braver than you believe, and stronger than you seem, and smarter than you think."

– A.A. Milne

Separation Made My Family Stronger

Staying strong in mind, body, and spirit was my challenge. Having inner strength, determination, and energy to get through the months of separation from my family unfortunately turned into two years. A total of 1,805 miles separated Arizona from Wisconsin. It would be the longest two years of my life.

My husband, Bill, and I have three boys, and while we lived in Arizona, we planned on moving back to Wisconsin to build on the land we had recently purchased. My sister and her husband purchased the land with us; we split the property. We both had 18 acres and decided we would live next to each other. My sister and I are originally from Wisconsin, and our husbands are from another state.

My sister and her husband moved back first. Bill helped them by driving one of their vehicles to Wisconsin. Then Bill flew back to Arizona to get ready for his move to Wisconsin; however, I wouldn't be returning with him just yet because I was in school. He would start his new job immediately.

A million thoughts went through my mind: *How long would we be separated from each other? How many boys were going to move with Bill? One, two, or all three boys? What about the boys' schools? Should we risk pulling them out early or wait until summer? Where were Bill and the boys going to live? We didn't want additional living costs. Would I finish college or dropout and move? What would the move date be? What about the cost of us living in different states? Groceries, my college payments, school lunches, daily bills, gas, etc? We had to be cautious and careful of our spending. Decisions and sacrifices had to be made regarding saving money and being apart to be together again. How would this affect the boys? How would this affect us as a family? Also, we needed to think about the house we currently lived in. What realtor were we going to use? When were we going to list the house on the market, and for how much? We had so many things to discuss not to mention the emotions, stresses, and frustrations we were feeling about this upcoming adventure. Do I call it an adventure? Or were we just plain crazy?*

When we finally decided on a plan, we had to have a conversation with the family members who would be helping us. We decided that Bill was going to move to Wisconsin with my oldest son, Keigan. I was going to stay in Arizona with our two youngest boys, Kaulen and Brenden. We also thought about Bill and Keigan temporarily living with my mom when they arrived in Wisconsin. She still lived in the very small house I grew up in. There wasn't a lot of room for them both even though it was a three-bedroom house. Bill living with my mom was a possibility, and Keigan could live with my sister. However, this would affect my mom and sister too.

Dropping out of college was not an option for me because I wanted to succeed in completing my classes which couldn't be taken online. Even though working full-time, attending classes part-time, and being separated from my family was going to be a struggle, I knew it was the only way.

We didn't know of the challenges that lie ahead involving my son living with my sister. *Would they be comfortable making sure he was getting his homework done, especially with the unknown factors of starting a new school? Would Keigan's grades suffer from us not being together as a family?* They would kind of be like substitute parents to him. Even though Bill wouldn't be far, he still wouldn't be living in the same household as Keigan.

My Dad passed away in 2004, which was only three years prior, so my mom lived alone. I wasn't sure how she would feel about Bill temporarily moving in. Maybe it would be good for her because he could help her around the house with things my dad would have worked on like yard work and physical labor. Also, just knowing that Bill would there to keep her company gave me a sense of happiness. Plus, I knew Bill wouldn't go hungry because my mother loved to cook.

After discussing our plan with both my sister and my mom, they agreed. They both knew that it would just be temporary, but at this point, we were not sure for how long. Regardless, they took that leap to help us, and I am forever grateful for their kindness and encouragement.

During the summer of 2007, when the big separation of our family began, the whirlwind of emotions set in. Keigan had just finished his freshman year. Kaulen had just finished eighth grade, and Brenden finished fifth grade.

Bill and Keigan would bring minimal items with them to WI. Mostly everything would stay with me in Arizona to prepare and pack. Also, I needed to get the house ready to be put on the market to sell.

The next two years, from the summer of 2007-2009, were not easy. All three boys had different feelings about what was about to happen with our family. They were open to it, but they also had feelings of sadness, anger, and sometimes didn't talk about it. They would miss their friends and at times didn't want to move.

During the difficult times, they struggled with their homework and turning it in on time. Kaulen and Brenden missed and skipped school on days when I was at work. I found out when I received a letter from the school indicating if the boys continued missing classes and school, the consequences would not be good. I was really upset and frustrated because I felt like I was a single parent even though I was married. I had a talk with the boys and told them we could all get in trouble for them ditching school. I didn't know what else to do at the time. This had never happened before, and I think they were trying to see what they could get away with.

I learned of this from one of the boy's letters that he wrote a paper on, for school of all places, of what they were doing while I was at work. They would go outside and play basketball in the driveway and go for bike rides. This was the kicker though! I have really long bamboo poles, and they would lay the poles on the stairs, and then slide down on their pillows. I had to laugh about that one thinking how fun that must have been. I also wonder, *What did the teacher think after reading it?* As a "single" parent, I had challenges with the boys. We did have a lot of good times though exploring new places and creating fun memories.

Keigan also struggled with turning his homework in on time. But overall, he had pretty good grades. It was great that he had his dad around to help with that. It is so ironic that Keigan ended up going to the high school I went to and graduated from. Keigan adjusted well with meeting new friends at his new school. He also joined the military and went to boot camp between his junior and senior years.

I believe that's what kept me going during the entire time I had to be strong—thinking about all the military men and woman who are separated from their families and fighting for the freedom of our country. Their circumstances were much tougher, and I knew I would be able to make it through. It also helped knowing that this separation was only temporary. I tried to stay positive, but there were some challenging times. I allowed myself to cry, and I prayed a lot too.

Some of my achievements I'm most proud of were successfully making the honor roll in college, as well as the Dean's list, along with becoming a member of the Phi Beta Kappa all while working, being a somewhat single parent, and getting the house ready to sell.

During the two years we were separated, Bill was only able to visit us in Arizona twice. Once, because I had to have surgery for a kidney stone removal and the other was to see me graduate from college. The two younger boys, and I did not go to Wisconsin at all during those two years because we were trying to save money.

We decided two years apart was long enough, and we needed to reunite our family. During this time, we put the house on the market. This is when the market crashed in Arizona; it was not a great time to be a seller. We found someone who would do a rent-to-own. We thought this was a great option because it was our only option! We finally moved to Wisconsin. Things were going great at first with the couple we were working with on our home. They paid their rent on time, but then, within the first year, the husband passed away. The following year, the wife decided to move out of state without telling us. I found out from a friend's notification comment on a networking site. We then contacted the wife and wanted answers. She told us she had to back out of the sale; the house was ours again. We ended up finally selling the house, but it was a short sale and we lost money. We also couldn't buy a house for three years. That's the downfall of a short sale on a house. We needed to rent, until that clause was no longer valid, before we could purchase a home. All that frustration added to the complications of what we were going through at the time.

After the two years, we were able to move to Wisconsin and once again all be together as a family under the same roof. It was a bit of an adjustment all living together again, but it worked out.

Sometimes life changes and takes us down a different pathway in a direction never intended. My sister and her husband ended up selling their portion of the land, and neither one of us ended up building a home on our land. So, the dream of building homes and being next door neighbors did not happen. My husband and I held onto our land until just a few years ago and then sold it.

Fast forward to current times, I literally am in the process of moving right now, so the flood of emotions is stronger than ever as I sit here and pack our belongings. This time however, it is a little different. This time Bill and I are leaving Wisconsin and our children behind. They are in their late 20s, and I am crying as I type this. As I go through what to pack, I am finding items that belonged to the boys as they were growing up. Like old schoolwork papers, baby blankets, toys, and some other miscellaneous items.

I am reflecting as I read some of their schoolwork papers and the stories they wrote; I will now pass this on to them. They can keep what they want, and I will definitely be keeping a few things for myself a little while longer too.

This is moving week, and the boys have been coming over to help load up the moving truck. Watching them load our stuff, I am overcome with a sadness; they are loading up their parents' stuff that will be moving across the country without them this time.

Well, the moving day is here, but not the original date; that was yesterday. The moving truck was not big enough, so we had to get an additional small trailer to be pulled behind the moving truck.

We are officially on the road. We are in the truck, pulling our car. Our friend is behind us driving the moving truck which is pulling a smaller trailer. We are paying for his flight back to Wisconsin next week. I am writing part of this chapter on my laptop as my husband is driving. It is really hard to see what I am typing through the tears of sadness, happiness, and excitement for this new adventure in a state that we have not lived in before. It is in the south, and we have been vacationing there for the last five years. We enjoy it there as well as the weather. We wanted to move somewhere warm but not too hot, and my husband did not want to move back to the desert.

Going through the moving process, reminds me of all the challenging times we went through, what we needed to do before we left, and the last-minute mishaps that happened. We went through difficult times during this process, but we didn't sweat the small stuff.

How I took back my power.

No matter what you are going through, just remember when you are feeling the mix of emotions, you are strong enough to go through it. Have confidence in yourself which will help guide you. Also, keeping your mind busy with work, reading, puzzles, or any activity helps. Getting out and exercising like walking or biking is great for your body. During the separation, I lost 40 pounds. If you are a spiritual person, praying and talking with friends will help tremendously. Be positive! If I'd had a negative attitude, the outcome would have been worse. You have to work for it, and only YOU have control of your inner power. You might not think you can get through it and wonder if what you're doing is the right choice, but when you are able to see what is beyond, it will be all worth it.

About Karen

Karen has been married for over 34 years and has three wonderful sons. Her boys are now adults, and she spends as much time with them as she can. She has lived in Wisconsin and Arizona most of her life, and although Wisconsin is beautiful, she prefers to live in a warmer climate. Karen has worked in business, theater, film, TV, and as an on-air Radio DJ under the stage name "Karen Makaye." Karen misses working on a film/TV set and hopes to be involved in more productions in the future. She loves being on a movie set and learning how a film is put together from beginning to end. This is Karen's first book. She has always wanted to write a book from the heart and based on her personal experiences, so this is a great step in that direction. She enjoys spending time with family and friends, especially around the holidays. When she was a child, she loved to go up-north, in Wisconsin and spend time with her family. She misses the days of spending time on the lake and fishing with her mom and dad. In her spare time, Karen loves to refurbish furniture. She also loves crafting and painting with friends. Karen enjoys traveling across the country, browsing and/or shopping, trying out different restaurants, and touring history destinations. She also enjoys the outdoors hiking, music festivals, and reading a great book on the deck with a cup of coffee.

Karen can be reached at karencarr111@hotmail.com

SCOTT MELESKY

"It is not the critic who counts; not the man who points out how the strong man stumbles, or where the doer of deeds could have done them better. The credit belongs to the man who is actually in the arena, whose face is marred by dust and sweat and blood; who strives valiantly; who errs, who comes short again and again, because there is no effort without error and shortcoming; but who does actually strive to do the deeds; who knows great enthusiasms, the great devotions; who spends himself in a worthy cause; who at the best knows in the end the triumph of high achievement, and who at the worst, if he fails, at least fails while daring greatly, so that his place shall never be with those cold and timid souls who neither know victory nor defeat. Shame on the man of cultivated taste who lets refinement to develop into fastidiousness that unfits him for doing the rough work of a workday world."

– Former U.S. President Theodore "Teddy" Roosevelt's
"Citizenship in a Republic Speech"
in Paris, France - April 23, 1910

Taking Back My Life
One Step at a Time

I have faced adversity in my 30 years as a sports journalist and educator. Job loss, accidents, and illnesses have all plagued me over the past three decades, but I've found ways to overcome each of these challenges and move forward to achieve individual and group success. On June 12, 2017, I would face my most difficult challenge in my life which I feared would ultimately end it.

The day started like any other. I was entering my seventh year as a first-grade teacher at the school I worked at. It was the beginning of the summer session, and I looked forward to the curriculum I had planned for the next three months. It would be a summer of field trips, crafts, and learning. I was very excited. The school had its annual kick-off barbeque lunch to open the session. The meal was cheeseburgers and M&Ms. It wasn't the healthiest fare, but it was comfort food to make everyone happy and excited for the days to come. I had my class line up for the food and drinks. When we were all settled, we sat in front of our classrooms on colorful festive blankets. I loved comfort food and eagerly ate my burger and candy. As I took a bite of the burger, I realized mine wasn't fully cooked. The M&Ms I ate were also old and expired. My stomach began to feel sick, and I felt nauseous and queasy. I drank some water and took Tums to settle my stomach, but I continued to feel worse. My wife picked me up from work, and I went to my doctor. I told him my symptoms, and he told me I had food poisoning and prescribed medication.

When we got home, I took the medicine and went to bed. An hour later, I woke with the most excruciating pain I ever felt. My legs swelled up immensely, and my muscles completely locked. I was immobile, and the pain was unbearable. My stomach swelled up to the size of a basketball, and I felt my insides expand. I began to vomit black liquid. I was scared, and honestly, I thought I was going to die.

My horrified wife called 911. The ambulance workers were professional, but I could tell from the look in their eyes I was in bad shape. One worker offered to put on Game 5 of the NBA Finals in the ambulance to try to calm me. I declined; I did not want to connect the NBA Finals to this horrid day because I am a big basketball fan. The ride was fast, but my stomach and legs still hurt immensely. It was surreal; I could not hear the sirens, but the two female medics were efficient and attentive. They asked me to rest in the ambulance as they took my blood pressure. It was dangerously high at 173 over 120. I was determined to remain calm and not panic, but I knew I was in for a long evening.

The ambulance reached the hospital, and the two medical workers gently put me on a rolling bed where I was greeted by two more doctors. My stomach expanded to three times its normal size during the ambulance ride, and the pain was excruciating. It felt like my insides were extending and trying to leave my body. With each extension, the pain grew exponentially. The doctors looked concerned and took my blood pressure again which was very high at 185 over 125. They quickly rushed me to the Intensive Care Unit (ICU). I was informed by a doctor that my intestines were inflamed, and my stomach was swollen; that's why I couldn't keep any food or drink down. The doctors then put me through endotracheal intubation which is the process of inserting a feeding tube into my mouth, so I could be fed intravenously. It was a very painful experience as I vomited three times before the insertion was successfully done. It hurt and felt awkward as I felt the tube go down my throat and into my stomach. It was long and leathery, and during my nine days in the ICU, it felt extremely uncomfortable. The nurses then extracted my blood and injected an Intravenous (IV) bag full of potassium. This bag would be my constant companion for the next nine days; it would also serve as my nutrition. During my ordeal, my legs cramped and completely locked due to the lack of potassium; again, the pain was excruciating. The potassium in the IV would replenish the lack of potassium in my body. My leg cramps slowly eased. The doctors also hooked me up to a machine that sucked the fluids from my stomach since it wasn't working properly.

I felt terrible and looked even worse, but I felt strength from the team of 10 doctors and a group of hourly ICU nurses who worked feverishly to help me get through my painful situation. Every hour, the nurse would draw and test my blood. I was attached to another machine that sucked blocked fluids out of my intestines. The nurses were constantly monitoring and informing the doctors of my condition. I felt helpless and depressed, but I never gave up hope I'd get through it. Over the next five days, I saw my body shrivel off 30 pounds as I continued to be fed intravenously. I was depressed as my muscles shrank to weak rubbery limbs right before my eyes. I began to wither away as I was not able to digest proteins and nutrients.

My throat continued to burn as the tube felt like rough sandpaper scratching its way toward my stomach.

On the fifth day in the hospital, the doctors decided I was in better condition, and they let me try and eat some food. They were concerned about how much weight I had lost and cautiously removed the feeding tube. I was allowed a small cup of water and a small scoop of yellow sherbet. After not being allowed to eat or drink for five days, the water and sherbet tasted absolutely divine. The coldness from both helped my parched and dried throat. The doctors seemed satisfied and said I would be released home the next day. I was excited and relieved—ecstatic that I would finally be going home and able to eat and drink again. It was much easier to rest now.

As I was sleeping, four hours after my meal, I was awakened by a searing burning sensation in my stomach. It felt like I was on fire from the inside again, and I could do nothing to extinguish the fiery pain. I screamed in pain and hit the alarm to get assistance from the nurse. She came quickly and checked my vitals. My stomach began swelling to the size of a beach ball, and the accompanying pain was horrible and beyond belief. The doctors immediately flurried in. They told me my stomach and intestines were swollen and agitated from the sherbet and water. My stomach could not digest the food and water, and my intestines couldn't break them down. There was no place for the food to go except back up through my stomach which created this searing sensation in my stomach and groin areas.

I was immediately rushed into surgery. I had two previous surgeries for a hernia—one as a child and one as a college student. Both surgeries had been done incorrectly which led to my swelling and inflamed intestines. This would be the third surgery I would have for a hernia in a span of 37 years. My intestines would have to be cut and shortened to prevent further hernia surgeries in the future.

I was always scared to "go under the knife" and hoped that I wouldn't have to when I was initially admitted to the hospital, but the incredible pain ripping through my stomach made the prospect of surgery a relief. The seriousness of the surgery gave me pause when I had to take off my wedding ring and give it to my wife. We both teared up, but I told her that I would "be back" to get the ring. It was the first time that I had taken off the ring since we were married back in 2014. The surgeons slowly rolled me into the operating room. I was in pain, but again I relied on my faith to get me through. I said a silent prayer of protection for myself as the anesthesiologist placed the mask over my nose and mouth; my mind and world went dark.

I woke up hours later in my hospital bed with no memory of the surgery that was performed on me hours earlier. Several blankets covered me as I shivered. My legs also cramped during the surgery, so a machine was gently moving my legs up and down in a marionette-like fashion to relieve the cramps and keep the blood circulating in my legs. To my relief, the pain had stopped, but the incision that was stapled through my belly button gave it a ghastly look. I asked my wife for my ring which she eagerly put back on my finger. "Is it over?" I asked.

She nodded yes, happily. I smiled weakly and went back to sleep, but my ordeal was not quite over yet.

The next morning, I looked at the incision from my surgery. The staples were sharp and clean and resembled the incisions on the Frankenstein monster; however, what stood out was the redness and green secretions that surrounded each incision. When the doctor inspected it, he became annoyed and then informed me and the nurses that the surgical incision was infected and had to be rectified immediately. He attached a small machine to my new surgical incision. The machine was designed to suck the infection out of my body and wound. Despite the wonderment of this medical technology, the process of the healing would be slow and methodical.

In addition, a breathing apparatus assisted me with my breathing throughout the day building up my lungs which were also weakened from my six days in the hospital. My legs had also deteriorated from six days in bed and the surgery. There was still a machine under my bed massaging my legs, but my doctor said that I had to do more.

"I want you to walk two hours a day, to get the muscles in your legs stronger," he said. "If you don't become more active, your legs will atrophy making it harder for you to get out of bed."

I agreed with my doctor but looked at my body which had become a complete mess after this horrid week. I lost 30 pounds of muscles in my arms and legs. I looked and felt like a deflated rubber toy. Two hours walking, when I was healthy, was easy for me. Two hours walking in my current state would be something completely different. My face fell when the doctor told me what was expected of me, but my wife put her hand on my shoulder lovingly and said, "We got this!"

For the next three days, I pushed my body and mind harder than ever before as I broke up my two-hour walk around the hospital into four half-hour increments. Getting out of my hospital bed was an exercise in itself. One nurse would help steady my IV, which was still in my right arm, and unplug the machine massaging my atrophied legs. Another nurse would help me get out of bed and assist me with my walker. I also had to cradle the machine that was now attached to my gut sucking the infection out of my body. I was weak and inactive and took a large breath from the breath machine before I walked to strengthen my weakened lungs. My first hospital walk would be agonizing. My body was frail at 160 pounds on my 5 foot-11 inch-frame. I was holding an extra five pounds between my infection cleaning machine, IV, and walker. Plus, I was already out of breath just getting out of bed.

The moment my feet touched the ground, I felt like I was taking back my life. I was taking back my life from injury and inactivity. I wanted to be able to work again, write again, and enjoy the highs of life. I wanted to do this for my wife. We had been married for three years, and my heart ached with what I was putting her through. She stayed every day by my bedside giving me the much-needed love and support to fight through the constant pain I was enduring. I wanted to do it for my family. My parents and siblings were in Connecticut and Japan. I did not want to alarm them and have them come to California. My wife called them every day per my request to give them updates on my health. I was overcome with emotion talking to them, but my wife was a "rock" for all of us. When, I took those first steps in the hospital room, I believed that I was taking back my life and was going to move forward full throttle mentally, physically, and spiritually to get it back.

When my feet hit the ground, my legs began to buckle from inactivity. I pushed into my walker for leverage and cradled my infection sucking machine like a newborn baby. My wife helped steady my IV, and we slowly walked out of my hospital room. My last memory of being outside my room was when I was taken into the ICU. I breathed deeply and started to tear up from those memories. I then heard a loud clapping noise and boisterous cheers in my direction. It took my breath away. The doctors and nurses were clapping and cheering me on—exhorting to continue my walk. I overheard a medical intern miff incredulously, "I can't believe that is the same guy we put in the ICU last week. He has come such a long way."

I slowly lifted my arm and clenched my fist in a pumping motion, and I smiled weakly to acknowledge the vocal and emotional support I received from the hospital staff. It really inspired me to work as hard as I could to overcome this challenge and succeed in my life. On this day, I was only able to walk to the end of the hallway and back. By my third day and 12th walk, I was able to walk to the end of the hospital motivated by the cheers and support from the hospital staff and my wife's family and friends who also were at the hospital every day to support me.

On the ninth day in the hospital, I was released. I was slightly stronger from constantly walking, and I was able to have my first meal: beef broth, coffee, apple juice, a turkey sandwich, and a lemon gelatin cube. After not being able to eat for nine days, this meal too was divine for me. It helped strengthen my body so I could take my last walk around the hospital. I hugged and shook hands with every doctor and nurse. One of my nurses named Angela created a pillow out of hospital sheets to alleviate the bumping that I would feel on the 30-minute drive home. I was wheel-chaired out of the hospital by a very energetic and peppy student intern who inquired if I liked sports. I told her that I was a sports journalist and a teacher which excited her very much. I then added that I was hospitalized during Game 5 of the NBA Finals and did not know who won the series. She seemed shocked and said "Golden State won 129-120 and beat LeBron (James) and the Cleveland Cavaliers in five games. I can't believe you missed it."

I wryly smiled and answered that "bigger things happened to me that day" (referring to my hospitalization). We both laughed at my dark humor, and we wished each other the best as my wife drove over to pick me up.

My wife gingerly helped me into the car. I cradled the homemade pillow the nurse made to help my stomach from hurting from the impacts of the road. I also held the machine that cleaned my infection. It would be my companion for the next two months. The pillow worked as I pushed it against my stomach when the wheels of the car hit a pothole or gravel on the streets. Once we got home, I carefully got into bed. I was exhausted from my nine-day hospitalization. The humming of the machine's mechanical healing lulled me to sleep.

The next two months would be geared toward getting back into physical, mental, and emotional shape to resume my teaching job. My third hernia surgery and swollen intestines and stomach ravaged my body and mind. I lost 35 pounds in nine days severely affecting the muscles in my arms and legs which critically impacted my mobility including walking and lifting. My lung capacity was also shorter and had to be increased dramatically in order to get me back into working shape. My diet was also reduced to no fiber and small amounts of protein as my stomach and intestines recovered from the surgery. Broth and small amounts of pork and oatmeal would be my staple diet. I had a strong support team around me. My devoted wife and a visiting nurse would be there to clean my wounds, discard the infectious germs from my cleansing machine, and help me with my breathing exercises to set up my comeback to take back everything I had lost in the previous nine days.

It was a strenuous daily schedule that I undertook to get myself healthy. I was determined to work hard every day to meet each goal. My morning would start when I would drain the fluids from the machine that was sucking the infectious germs from my surgical wounds. It was tedious and disgusting, but I knew my body was cleansing and healing, so I did it each day without complaint. The visiting nurse would then come by and clean the wound and discard the infectious germs. My wife would then make me a breakfast of tomato broth and Gatorade. The nutrients from both fluids would energize me. I would then use the breathing apparatus the hospital provided me and would blow into for 10 minutes, 10 times a day. I would have to breathe until the flaps in the apparatus stayed open. It was a hard exercise, and I did have trouble with it early on in my recovery. My physical recovery included walking around my apartment for two hours (in 10-minute increments). I would be lifting one-pound dumbbells twice a day for five minutes to build up my arm strength. The rest of the day I would rest my mind and body.

Though, I had tremendous support from my wife, family, and hospital staff, I would get depressed at the slow speed of my recovery. Fortunately, one man who helped me get through my dark times was former United States Marine Corps member and World War II veteran, the late Sgt. Edgar Harrell. (He would pass away at 96 in 2021). He was one of six living members who survived the U.S.S. Indianapolis sinking. The ship was on a classified mission delivering enriched uranium for the formation of the first atomic bomb located on the island of Tinian. After successfully delivering the uranium to the island, the ship was attacked by an I-58 Japanese submarine. It hit the U.S.S. Indianapolis, and the ship sank in 12 minutes. Harrell was one of only 316 crewmen who survived out of 1,195. There were also 879 crewmen who were either killed by the blast or sharks over the next five days that they spent in the ocean. Harrell survived this horrible ordeal and had a positive impact on people by sharing his faith. He also wrote a book *Out of the Depths* in 2018 about his experience aboard the U.S.S. Indianapolis.

As I struggled with my recovery, I read his book and was inspired by his strength, courage, and determination to overcome his horrible ordeal. I was inspired and worked hard on my rehabilitation to get back into tip-top physical, emotional, and spiritual shape. I emailed him, and we exchanged correspondences. I thanked him for his service and sacrifice, and I told him how his life had inspired me to work through my own physical setback. He thanked me for reading his book and told me not to give up and to keep working hard which I did. He would send me eight autographed pictures along with his words of wisdom which were key in helping me take back my life.

By the end of my first month home from the hospital, I had gained 10 pounds of muscle from the broth and protein I consumed. My lung capacity increased significantly, and the two hours of walking and lifting helped my stamina and raised my energy levels. I was still using the infection cleansing machine, but I was now taking slow walks around the block. I kept believing in myself and knowing that with determination, hard work, and a positive attitude I would take back my power.

I kept on working on my mind, body, and spirit each day, and I returned to my teaching job in August. I would continue to teach for another three years and earned my master's degree in education in two years. I am now currently an author who has contributed to eight different books. Back in 2017, when I was hospitalized with a hernia, I never thought I would be able to overcome it and reach the successes that I have. In order to succeed and take back my life, I never gave up on myself. I believed and worked hard mentally, physically, and spiritually.

How did I take back my power?

- I got my physical strength back by walking every day.
- I got my health back from eating healthy foods and building up lung strength with a breathing machine I used.
- I got my mind right by reading plenty of inspirational and spiritual books.
- I got my spirit back through plenty of prayer.
- I got my writing skills and academic skills back by going back to school to get my master's degree in education and have been published in eight different books.
- I never gave up on myself and always believed that I would get though this by staying strong and positive, as I continued to move forward with the loving support of my wife.

About Scott

Scott Melesky has been a sports journalist for over 30 years. He graduated from Syracuse University with a bachelor's degree in history in 1995. Melesky earned his master's degree in education from Pacific Oaks College in May 2021 and worked on the college's research team in the School of Education from 2021-25.

Melesky was a guest speaker at the 2013 Glendale Community College's Early Childhood Education Conference, the 2023 Green Dot Schools Legacy Foundation Conference, the 2024 UCLA Center of Transformation of Schools, Pacific Oaks College, and Pasadena Unified School District's Our Children Can't Wait: Ensuring Equity For All Conference, and has been a guest on Mark Mancini's Breakin It Down podcast and Maryann Castello's Liberty Bell Smack on WWDB AM 860 Philadelphia in 2023.

He has contributed writings to the Society for American Baseball Research's books: *Yankee Stadium 1923-2008: America's First Modern Ballpark, Nichebi Yakyu: U.S. Tours of Japan Vol. II 1960-2019, Sox Bid Curse Farewell: The 2004 Boston Red Sox, Yagu: Past, Present, and Future of Korean Baseball, Dr. Jorge Iber and Anthony Salazar's Beisbol on The Air: Essays on Major League Spanish-Language Broadcasters,* and James Cryns' *On Story Parkway: Remembering County Stadium.*

Melesky has also contributed to Professional Football Research Association's 1976 Oakland Raiders, Nick Del Calzo's *My Baseball Story: The Game's Influence on America,* David Krell's *The Fenway Effect: A Cultural History of the Boston Red Sox.*

He has worked for and contributed stories as a Sports Editor and writer for the Los Angeles Daily News, The Glendale News Press, The Patriot Ledger, The Syracuse Herald Journal, The New Haven Register, The Waterbury Republican, The Naugatuck News, The Southington Observer, The Westchester Daily Local News, The Van Nuys News, YardBarker, L.A. Parent, Highlights, The Coffin Corner, Japan Ball, Team Fenom, The Sports Page, SPM.com, and YWCA Greater Los Angeles. He also contributed support staff game work for ESPN, MSG Network, and NHK World Japan.

Melesky has also worked in the sports information departments at Marquette University, Ohio University, Quinnipiac University, and the Florida Institute of Technology. He is a member of the Society for American Baseball Research, Institute for Baseball Studies, Association for Professional Basketball Research, and Professional Football Researchers Association.

CONNIE F. SEXAUER

"I want to be a woman who overcomes obstacles by tackling them in faith instead of tiptoeing around them in fear."

– Renee Swope

Reclaiming My Life

How to take back power? A major first step is recognizing one has lost power. This moment is compelling, and the correct tools are needed to move forward. I have faced difficulty many times in my life, and rarely do I automatically respond with a clear solution.

In the case of my 25-year marriage it took over 24 years to face the realization that not only was it over, but it probably should have never taken place. I fought a battle of denial as I so wanted to succeed and not prove I was a failure.

In the case of battling breast cancer, I had to face a dreaded diagnosis and seek help to move onto a path toward survival.

When I walked the path of helping my son through a severe car accident plus the repercussions of that ordeal, I had to let go of control, put my faith in a higher power, and figure out a survival plan for both of us.

Thankfully by the time the COVID-19 Pandemic hit in 2020, I had these past experiences to rely on. However, I was not prepared for the health problems that would follow and what it would take to find my way back to normalcy.

I did not contract the pandemic, but the fear I faced that I would get the disease and eventually die from that illness placed me in a deep state of depression. In the spring of 2020, I was living a good life of fun and contentment. I had retired in spring of 2019, and the following fall I had published two books. I scheduled a national book tour to discuss my work. It was an exciting time that I embraced with enthusiasm. One of the books I had worked on for over 20 years. The topic was a passion of mine—the history of the St. Louis Cardinals playing fields. I am an avid baseball fan and looked forward to presenting my work and fielding questions. I was in Florida attending spring training and promoting my work. Then out of nowhere the spring training season was cancelled midway through due to the threat of the spread of COVID-19 nationwide. The record number of cases worldwide and within the United States required folks to take stock of the circumstances and react by simply dialing back. I got the message loud and clear. The next day I headed home to Wisconsin to hunker down and seek a safe asylum.

At the time I had no idea how this fear would overtake me and cause me to go into a severe depression. I traveled all the time. I had been a college professor, and once I turned my grades in each semester, I took off to visit friends and relatives or conduct research nationwide. Travel excited me and kept me in touch with my emotional side of life. I figured the shutdown of the country would last a few weeks and all would be back to normal in short order. I was wrong. Not just wrong, but far off-base. My fear of getting COVID-19 and dying kept me isolated from even doing things like walking daily, going to the movie theater, shopping, and heading out to dinner with friends. I lived alone, and I now sheltered in place. I didn't go anywhere or do anything except watch television 24-7 which broadcast the horrific stats that proved the world was being overtaken by a disease of huge proportions.

Summer of 2021, I was invited to a family wedding in Indiana. I was ecstatic to have a reason to travel and visit friends and family. I had no idea that my denied depression would finally catch up with me, however; I fell three times on that trip. Others noticed and made the decision to take me to the emergency room to see what the problem was. I was not happy about that. I thought I was fine and was simply tripping over unfamiliar territory. The doctor decided I had a urinary tract infection and lacked sodium. He suggested I head home and see my primary doctor for further testing. I made a quick stop to visit a friend in Chicago and then returned to Wisconsin. My primary doctor ran additional tests but could not pin-point any immediate problems. I thought all would be fine. It wasn't. Within the next several weeks I fell quite a few times. A dear friend, Pat came to visit in August, and I fell in front of her. She was astonished at how it suddenly came over me and without any warning, I just went down. I didn't get weak in the knees or show any indication I was going down. I returned to the doctor, but she was clueless as to why I would fall. More tests were scheduled, but nothing showed an abnormality. Meanwhile, I was faced with a major life changing decision. I decided to move to another town in Wisconsin which was about 40 miles away; it was where my divorced son lived. He had custody of his young children every other week, and I thought I could help him if I lived closer. My condo sold in a short time, and before I knew it, I uprooted myself from my surroundings and began a new journey. In the meantime, I continued to fall without warning or reason. I went to the doctor again and even to the emergency room. There seemed to be no logic as to the reason I was not stable on my feet. Tests showed nothing out of the ordinary.

Thankfully, a dear friend, Vickie, helped me coordinate my move; she was the first person to point out to my son that I was in bad shape. Over my lifetime, I had moved more than a dozen times and I was good at taking control of the move. Not this time. I didn't lift a finger to help. I didn't pack a box or load a car. I hired movers but still needed help with getting everything together. My friend called in help from other friends and local family members. She too was baffled at my lackadaisical attitude. I simply would sit and stare while others worked diligently around me handling my personal possessions. One day she and my son, while loading up one of my closets, had a heart-to-heart conversation. She said that he had to tell my daughters, both who lived out of town, just how badly I was doing. He agreed. He called his sisters, and they flew to Wisconsin to check it out for themselves. They took me to the doctor and questioned my behavior, but to no avail no one determined why I was so unresponsive to life. I moved to the new town, and my son would check in on me. I continued to fall. I requested an appointment with a psychologist. Perhaps I was going through a depression because my life had changed so drastically. The first psychologist I went to was a woman, about twice my age. She was no help, and we had no rapport. I requested another referral and was sent to see a male psychologist. He supported my theory of going through a depression but reasoned that it was only logical as my entire life had changed. His advice was to get back on a scheduled plan, and all would pass. That was a correct plan but too simple. With no follow-up, I simply continued my new ways with no improvement.

In January, Vickie suggested that perhaps what I needed was to hit the road again. She had a time share in Arizona, and one of her sons had moved to California. The weekend before we were to leave, I fell again. My son took me to the emergency room, and it was a resident doctor there who prescribed this: "I should not cancel my trip, but I must embrace it." He believed I was going through a depression and was dealing with malnutrition as well. We decided to take the trip, and I will say it was good for me. Not a cure, but I began to see improvement in my overall attitude.

While in California, I went to UCLA emergency room as I was suffering severe back pain and wasn't sleeping. The doctor there questioned me intensely and concurred with the depression and malnutrition. Though he didn't believe I was falling but fainting. He also had diagnosed a urinary tract infection. Antibiotics cleared that up in short order. A breakthrough finally occurred with this visit to the emergency room.

While I was there, they had me walk the halls on four separate occasions. Before I was released, the nurse and I started the walking route, and we stopped by a brightly lit room that had plenty of fresh sunshine—good for Vitamin D and reading material. Well due to the threat of COVID-19, the reading material had been stripped; however, there was a four-year-old *Readers Digest*. I took that back to the room with me. I began by reading the jokes and then found interesting articles. While reading a rather long joke, I suddenly burst into tears. It dawned on me I had not read a thing in over a year. I was an avid reader since age 10, and yet for the past two years I hadn't read anything. I would buy the newspaper and flip through it and then toss it in the trash. I read nothing, not even the magazines I subscribed to or to edit my writing. The realization that I had given up this pleasure in life was eye-opening. I started to think more analytically and realized I was so depressed that I didn't recognize my own depression.

While I lived alone, others had no idea about everything I had given up. I became malnourished simply by not having a desire to eat. I was fainting and not merely falling down. With this new awareness, the doctor was able to find a path to help me get my life back. He gave me information on a balanced diet and started me back slowly on a new exercise regime of daily walking. As he pointed out to me, my daily walks of six miles a day were more than exercise. They were my connection to life. When one takes a walk, they take in their surroundings and interact with those they meet along the way. I had lost that interaction with the world. Not only did the reading help me to think again about what I was reading, but it helped me regain my thinking skills. I was a listed person who planned their day and got things done. I had lost that habit as well. Once I began to think again, I started to think about my life and began to plan my path to success.

I was back to caring about taking care of myself. Over the next months, the restrictions on the country lifted, and things opened up again. One evening, I attended a musical presentation and began to cry. I cried because I realized all I had missed through my isolation. Music, plays, films, reading, attending sporting events, interacting with others all filled my soul and enriched my life. The pandemic and my depression took me away from everything and everyone I love. I cried because I had found my way back to living. Not only did I return to reading, but I was back to listening to the radio, watching positive things on the television, and listening to my music. Days later, I attended a movie in a theater and found myself laughing out loud. I cried again. I couldn't remember the last time I had laughed. Laughter fills the soul with joy.

I have written this piece because I realize if I can be taken down by depression, anyone can. I never gave much credence to depression before because I felt that one controlled their own life. Well as you can tell, I firmly believe we have control over our mental state if we are in a healthy frame of mind; but we also need to first realize we are depressed. We cannot fix what we don't acknowledge. I was not on this depressive path alone. As I share my story, so many have shared what they and their loved ones went through during those horrific shutdown years. I know people from birth to death were affected emotionally by this pandemic and equally by the fear of the pandemic. I hope I have helped you realize if you experienced this you were not alone, and I pray you find the strength when challenged to find your path back to a balanced and fulfilled life. Always remember you hold the power, use it wisely.

About Connie

Dr. Sexauer taught U.S. History and Gender Studies at the University of Wisconsin–Marathon County for 16 years. She was a graduate of the University of Cincinnati with a specialty in urban history. The main focus of her primary research was the role of Catholics in the 20th century and the importance of faith in their lives. Her works include articles on Catholic Civil Rights action, and the study of Charles V. Vatterott, Jr., a St. Louis real estate developer during post-World War II. She argued that "faith makes a difference in a person's life." The subject of her book, *From a Park, To a Stadium, To a Little Piece of Heaven* brings together her love of history, cultural studies, social change, and changes in sports. This book was released summer 2019. Dr. Sexauer also published articles and delivered national papers on these exciting topics of research. She was also a coauthor and contributor to the #1 bestselling book *Manifesting Your Dreams*.

ANN NEWMAN

"Connecting the dots of our lives, especially the ones we would rather erase or skip over, requires equal parts self-love and curiosity."

– Brené Brown

Unleash My Energy –
The Answer Lies in the Question

There is immense power in a question. I'm not necessarily referring to the existential questions about life and all existence, but rather, the exchange of energy when a question evokes an emotional reaction promoted by a simple question, such as "Why is that important to you?" Powerful questions can encourage us to reach into our souls to surface vulnerable ideas and emotions. That is, of course, if we choose to ask, and ultimately answer the question. I chose to reach inside my soul and ask a provocative question, "What does it mean to take back my power?" The answer surprised me and led me through a healing journey, and I hope it does the same for you. Here's my first question. Why do you feel you have to take back your power?

What is power? What is energy? Why does it matter?

There is a unique relationship between power and energy. The simplest explanation is power is the rate at which energy is transferred and energy is the capacity to cause change. But why does it matter? Each of us has the unique ability to utilize power to transfer energy. This energy can then be shifted to others or even within our own inner being. Remember, energy is the capacity to cause change.

Imagine you are hiking with a group of your best friends. You've been out for several hours and everyone is getting tired. You stop for a moment to adjust the fit of your shoe and then you notice you are alone and lost. There is no one around. You don't know where you are, where to go, or even how this happened. Your energy has certainly changed. Maybe your heart is racing, perhaps your mind is running multiple scenarios, or it could be that you suddenly realize how thirsty you are. You reach for your water bottle hoping a little hydration will help you think more clearly. You pull the light bottle up to your mouth and place the spout to your lips when only a few drops trickle onto your tongue. You are out of water. How do you feel? Powerless? Is your adrenaline fueling your energy? This scenario paints a dichotomy of power and energy. You decide the best choice is to close your eyes and try to think of the best next move. Then, you feel it. Someone is standing next to you. Startled, you flinch opening your eyes to see your very best friend gently touching your arm, asking, "Are you okay?"

You sigh and say, "I am now." This is also power and energy at work. We are often oblivious to power and energy's role in our lives. My job is to help you unveil what lies behind your power and energy, just like I was able to discover mine.

Before contemplating this chapter and mapping out the essential questions, I had never really given much thought to my own power and energy. I don't particularly think of myself as a powerful person nor do I spend much time thinking about the power I have, real or perceived. I was also blind to the energy I carry, the light I emit without even knowing it. Then I realized I wasn't looking at power or energy in the ways I needed to.

If power is a transfer of energy, the questions I needed to ask myself were these:

When am I transferring energy?

When am I not transferring energy?

How am I transferring energy?

Why do I feel like I've lost the ability to do that?

Loss. That's when the realization began to hit home. I was feeling a sense of loss. It didn't happen all at once. It was more gradual, like the plushness of new carpet that over time just gets worn down. You can see the tread of the places you walk more than anywhere else. My power diminished not from anything anyone did or took from me, but rather from the places I chose to walk again and again. It was time for me to take a different path, take back my power, and leverage my capacity to impact change. In other words, my power came from reigniting my energy.

How did I get my power back through energy?

I aspire to inspire and ultimately influence people to exert their passions. I never really thought about this as power. However, through my journey to reclaim my power, I have a new appreciation for my power as the ability to harness the energy that I create to inspire others. Sharing my energy has had a powerful outcome; one I had never imagined. Taking back my power is recognizing the gift of my energy to serve others, ultimately shaping ordinary moments and people into extraordinary occasions and amazing human beings. This is a personal power that is subtle yet profound—the kind that comes from understanding oneself deeply and embracing the unique energy that one carries. You have that same power and energy if you choose to ask yourself the right questions to find it.

Why did I lose my power? Where did my energy go?

A shout-out to all my former science teachers for helping me understand the law of conservation of energy, also known as the first law of thermodynamics, which states that energy can neither be created nor destroyed, only converted from one form of energy to another. My energy went somewhere else. Inward. Throughout the years, I have become increasingly more introverted. This doesn't mean that I don't talk to people, don't speak my truth, or engage with groups. It simply means that it takes an incredible amount of energy for me to interact with strangers, speak out in groups, and lead as an outspoken individual.

My good friends and family are surprised when I tell them I see myself as an introvert and that it takes so much energy for me to be outgoing. What they don't see is the inner drain to be present, curious, and aware of everyone else around me. It's that conversion of energy from channeling everything I have inside to create a positive impact on them. Don't get me wrong, I'm happy with who I am. In fact, I love that my power is quiet, hidden, and even preserved for those who are genuinely interested in taking the time to get to know the real me.

One day, I was presenting to a group of people. I had never met these people, yet my job that day was to help them understand how recognizing their unique talents, abilities, and motivations could have a positive impact on their career trajectory. I was there to help them see a new future professionally if they were interested. When I was finished with my presentation, one participant approached me and shared how my words touched her deeply. Through streams of tears, she proceeded to say how what I said spoke directly to her heart and that she had asked God to send her a message about what her next career path should be. My email inviting her to participate in the program was her first sign. Then, she said there was something unique and special about the way my words carried the "power of my energy" to transform her from not just knowing, but believing that she was destined for greatness and that it was her time to follow what she believed she should be doing professionally. She touched my shoulder and I could feel the magnitude of our collective energy. Then, for reasons I may never understand, the script flipped and she looked me in the eye and shared how she knew the power of my energy. She told me how I have a gift to touch hearts and minds with my words and the energy I carry with me, much like a light that emits with my presence. She also said that she knew from the moment she met me that although I present professionally as an exuberant personality, she knew that by the end of the day, I was drained by transferring my inner energy to others. I began to cry.

Strange, right? Here I am, the presenter, leaking with emotion beside a program participant. *How did she know so much about me without knowing me?* The truth is, at that moment, she knew me better than I knew myself. Through an immense energy, something powerful had happened between us.

I have a better understanding now that my energy is channeled to thinking, reflecting, being curious, and doing more that is less visible to those around me. In a world interrupted by extroverts, the struggle to fit in often left me feeling drained, disconnected, and often misunderstood for not having a voice or an idea. I was worn out from exerting energy that didn't feed my soul, ultimately, draining me and causing me to turn off my power. Then, I realized my power reignites when I can channel my introverted energy, in small ways, to other people. That transfer of energy, either through ideas, shared passions, excitement, conversations, listening, affirmations, empathy, or even appreciation, that's where the power is hiding.

Why is curiosity powerful? How does power connect to self-awareness?

I've always been curious about curiosity. Why else would I put curiosity at the center of my doctoral dissertation? There is immense power in curiosity because through curiosity we can explore various interests, passions, wonders, thrills, risks, and intuitions. It's fascinating to imagine everything out there that is left to be discovered, known, and explored. Yet, one of the most challenging places to be curious is about ourselves. In my pursuit to understand my own power, I discovered I needed to understand myself and spend time being more aware of why I let that power go.

It was in the quiet moments of reflection that I began to realize how essential curiosity was in my personal journey. The same curiosity that drove me to explore new subjects and meet new people could also be directed inward. I started asking myself questions that were both simple and profound: *What brings me joy? What drains me? What makes me feel alive?*

I discovered that my energy was a delicate balance of introspection and interaction. As an introvert, my natural inclination was to retreat and recharge in solitude. Yet, I also recognized that my energy had a unique quality—a vibrancy that could light up a room when I allowed it to. Curiosity helped me uncover this hidden potential. By exploring my internal landscape with the same enthusiasm I applied to external discoveries, I began to understand the nuances of my energy and how to harness it effectively.

How does an introvert muster the courage to shine?

Understanding my energy was one thing, but the real challenge lie in the courage to let it shine. As an introvert, there was an inherent fear of overexposure—a concern that showing my true self might lead to exhaustion or rejection. However, I realized that true courage wasn't about suppressing my fears but embracing them and acting despite them.

One of the most significant breakthroughs came when I decided to share my insights and experiences with a coaching partner—not an athletic coach. An intellectual soul coach. One of my people. It was an unassuming conversation. (That's how some of the best ones get started.) She asked, "So, tell me, what's on your mind?"

I shared that I was thinking about writing a chapter for a book about taking back my power, but I was struggling to have the jumble of ideas come together. Then she asked the most powerful question that changed my energy, "Why is that important to you?"

I was screaming inside: *Why is it important?! It's important because I have a story to tell about the passion and desire I have to serve people through what I believe I do well to create a powerful impact on them.* Most humbly, I am not the loudest, the quickest to share an idea, not the flashiest or most connected person in any room, yet, through my quiet energy, I carry a light of love and care and joy and genuine commitment to serving people for their own good. I want to find a way to unwrap this gift and deliver it to everyone. That's why it's important to me, and that's why I am taking back my power.

For years, I had kept my thoughts and feelings largely to myself, fearing that exposing them might be too draining or that others might not understand. As I spoke, I felt a mixture of vulnerability and exhilaration. The process of sharing my experiences and reflections allowed me to see my energy from a new perspective. I saw how my words resonated with others and how my authenticity invited connection and understanding. This act of courage did not deplete me; instead, it replenished me in unexpected ways. I realized that true power came from embracing my vulnerability and allowing my energy to flow naturally.

How is taking back my power and unleashing my energy a gift to others?

Reflection is an integral part of my journey to reclaim my power. It was through reflecting on my experiences, both positive and challenging, that I began to piece together the puzzle of my energy. Reflection was not merely about looking back but about understanding and integrating those experiences into a coherent sense of self. Myself.

Throughout my journey, one of the most important realizations was that my energy was not just for my benefit but could also be a gift to others. I began to see how my presence could positively impact those around me when I was mindful of how and when to share my energy. Thinking back to that powerful day with the program participant when we shared power and energy, I reflect on what was different. It's tough to capture the exact details, yet I know I was different. I was open, vulnerable, uninitiated, humble, and genuine. I entered the room, shared my passions, spoke from my heart, and knew my purpose was to give all that I had to those listening. At first glance, this doesn't seem different than how I typically engage, yet the nuance is that I trusted my inner power to guide my words and actions. I believed in myself. I affirmed that my gifts are worthy. Not in a boastful way, but rather from a place of confidence and assurance. I showed up as my whole self, rather than the person I thought I needed to be because I believed, probably subconsciously, that my energy fuels others.

Over time, in conversations with friends and family, I noticed how my thoughtful listening and empathetic responses created a safe space for them to express themselves. This realization was empowering. It highlighted that my introverted nature which I had once viewed as a limitation, was actually a source of strength. By channeling my energy with intention and care, I could offer a unique and valuable contribution to my relationships and communities.

Seeing the positive impact of my efforts reinforced my belief that my energy was indeed a gift—a gift that could inspire and uplift others.

Where is my ongoing journey leading?

Reclaiming my power and channeling my energy has been a continuous journey rather than a destination. It's a process that involves ongoing curiosity, courage, and reflection. Each day presents new opportunities to understand myself better and to manage my energy in ways that align with my values and goals.

I have learned to approach each day with a sense of intention. I now actively seek out moments of solitude to recharge, and I also embrace opportunities to connect with others when I feel energized. I have found a balance that allows me to honor my introverted nature while still sharing my light with the world.

Reflecting on this journey, I am filled with a sense of gratitude. The process of channeling my energy has not only empowered me but also enriched my interactions with others. It has taught me that power is not about dominance or control but about understanding and embracing one's unique qualities.

As I finish this chapter, I feel a deep sense of peace. The world around me may be loud, boisterous, chaotic, an extrovert's playground, but within me, there is a calm, powerful light. I have learned to harness this light with curiosity, courage, and reflection. It is a light that I am proud to share, knowing that it can bring warmth and inspiration to others. My light is my power and my energy.

In the end, reclaiming my power was not about transforming who I am but about discovering and embracing the energy that was always within me by asking myself the right questions. It is a gift, both for myself and for those around me, and it is a gift that I will continue to cherish and cultivate. The answer lies in the question.

About Ann

Dr. Ann Newman is passionate, curious, and driven to make a meaningful impact through small yet powerful actions. With years of experience as a trusted leader and thought partner in various organizations, she drives positive change and implements innovative solutions through the power of coaching and collaboration.

Ann cherishes time with her family, traveling to new destinations, and embracing the thrill of snowmobiling. A lover of good coffee, she finds joy in singing and listening to music. Guided by God's love and grace, she appreciates the beauty of life and stives to make each and every moment matter. Residing in beautiful Rochester, Wisconsin, Ann inspires others with her unwavering commitment to help them see their strength from within.

Doctor of Philosophy in Leadership for the Advancement of Learning and Service Dissertation title, "A Journey Down the Path of Curiosity: Exploring the Crossroads of Motivation Leading to Innovation," ProQuest Dissertations Publishing, 2018, 10812665.

Contact Ann: linkedin.com/in/annmarienewman

MARLA McKENNA

"I can do all things through Christ who strengthens me."

– Philippians 4:13

The Promise of God is Always Backed by the Power of God

When I think about taking back my power, I have to ask myself: *How did I lose my power in the first place?*

"For I know the plans I have for you," declares the Lord, "plans to prosper you and not to harm you, plans to give you hope and a future" (Jeremiah 29:11).

Pain, anger, depression, sadness, betrayal, and loss can often strip us of our sense of power. These emotions led me to believe I had lost mine. In the midst of suffering—self-loss, family-loss, friend-loss, and home-loss—I felt as though I had lost everything, including my power. There were times when I even questioned my faith. Yet, through it all, God offers His unwavering promise—a promise I needed to remember.

"My power will rest on you when you are weak" (2 Corinthians 12:9).

About 10 years ago, I made the difficult decision to go through a divorce. It's not something I would ever recommend, but then again, marrying the wrong person isn't something I'd recommend either. When you take your marriage vows, you never imagine that 18 years down the road, you'll face a divorce.

That said, my 18 years of marriage weren't all bad. In fact, there were many happy and wonderful times. The greatest blessings and accomplishments to come from that chapter in my life are my two amazing and beautiful daughters, Julia and Ashley. They were, are, and always will be the brightest lights in my life.

✶✶✶

"But he wants sole custody," I heard my attorney say after I told him I wanted to put the divorce on hold.

"He wants what?" I answered in shocking disbelief questioning my previous decision of wanting to try again.

As I hung up the phone, a chill washed over me; a storm was coming, and there was no turning back. The sky seemed to close in, and a wave of nausea hit me, leaving me gasping for air.

The past few months had been challenging, and the next several months would be nothing short of grueling. At that moment, it felt like I had nothing left to live for if I lost my girls. I wanted a divorce from my husband, not from my daughters. There is a difference. The waves grew higher as well as the rage within my soul. I became very anxious as I felt the strong current pull me under.

"Peace, be still" (Mark 4:39).

While it's reassuring to see the storm around us calm down, what matters even more is that our hearts remain at peace in the midst of the storm. While God hasn't promised to remove every storm from our lives, He has promised to give us peace through them, as long as we trust Him to care for us. I needed to trust God.

"Cast all your anxiety on Him because He cares for you" (1 Peter 5:7).

Divorce can portray an ugliness, and unfortunately my daughters were given a front-row seat to the chaos. Others, too, were drawn into the drama, watching, and commenting on this feature presentation— my divorce. Some people eagerly accepted their viewing invitations, some didn't, while others pushed for backstage access wanting to peer behind the scenes. It felt as if I were living in a reality TV show, with my personal life exposed for everyone to see. In my eyes, divorce should be a private matter between the two people involved—both taking responsibility for their actions, good and bad. It was a very dark time for me, and I felt like I had lost my power.

I was a wife; yes, but I was and still am just a human being—a mom, a daughter, a friend, a sister, an aunt, a godmother, an entrepreneur, an employee, a writer, a teacher, and a woman. I was a woman who had lost myself in a marriage that wasn't good for me anymore—a marriage that I needed to find the strength to leave.

Self-loss

It can be confusing when you lose your sense of self. Self-worth, self-esteem, self-direction, and self-growth can all take a significant hit, ultimately impacting your day-to-day life and sense of purpose. For a long time, I felt like I was trapped in my own skin—overwhelmed by anxiety, fear, unhappiness, as well as feeling alone. It was an unhealthy way to live. I had thought about divorce for a couple of years before finally deciding to pull the trigger.

Where would I even go?

How could I possibly afford to live?

How would I do this?

This was all I knew.

These were the questions that plagued me, and without answers, fear kept me paralyzed and delayed my decision to move forward.

"Fear not, for I am with you" (Isaiah 41:10).

How I found myself again. I immersed myself in the joys of my life.

I finally learned to put myself first. Some might call that selfish, but the truth is, you can't pour from an empty cup. You can't light the way for others when your own light within is dim. Over the last 10 years, I've learned to love me and take care of me.

I continued to pray every day, asking God to guide me and show me the path He had laid before me. While working on myself, I focused on growing internally, so I could find true happiness within and be able to share it with others. When my daughters and I were together, I made the most of every moment, creating new memories and reinventing old traditions. And when they weren't with me, I dedicated myself to my work and continued my personal growth.

- I leaned on my family.
- I built strong relationships with my friends.
- I poured myself into creativity.
- I wrote more books.
- I helped others.
- I made children smile.
- I learned how to manifest.
- I forgave.
- I found my self-worth.
- And I loved myself.

Find what brings you joy, and go there.

I love going to concerts, and the devastating phone call from my attorney that rocked my world and sent me spinning into the depths of a whirlwind, also prompted me to respond by seeking immediate joy. That joy always comes from going to concerts. While I feared I might somehow lose that joy as a result of the divorce, in that moment I was not willing to lose my power or my joy.

Rick Springfield concerts are a wonderful source of strength, joy, and unconditional happiness for me. Rick has become a friend. (Read my book *Manifesting Your Dreams* for that full story.) I looked at his tour schedule, and the concerts that weekend were in Pennsylvania and Maryland. Neither were drivable, but I gravitated toward the Maryland show. Last-minute Pennsylvania flights were very expensive, but I could get a flight to Maryland for $88. Yes, $88!

I checked for concert tickets. They were sold out; however, I reached out to Rick's management and was told two tickets would be waiting for me at will call. That was absolutely incredible. Rick has never let me down. I called my close friend, Penny, who is more like my sister, and said, "We're going to Maryland this weekend." She knew I wasn't in the best mindset, and this much-needed trip would be good for my soul. And it was. For 48 hours, I could just be Marla—a woman going to a concert with my friend—not a woman going through a divorce.

The weekend was truly nourishing for my soul, and it renewed my strength to move forward. The venue's intimate setting fostered a sense of sisterhood with long-distance friends I hadn't seen in awhile, and it deepened my connection to Rick's music—something that has always been a source of friendship and comfort through both good and bad times in my life. After the show, I was able to share some of my struggles with Rick, and, as always, he responded with kindness, encouraging me to take care of myself.

As I've said, Rick's music has always played a positive role in my life. Recently, he released his new album *Automatic*. I can't help but feel that it's no coincidence for me that it dropped right in the middle of my writing this chapter. When I heard the lyrics to "This Town," by Rick Springfield, a few of his inspirational words instantly transported me back to that weekend in Maryland—when I had to escape my life for a few days.

"This Town"

Yeah, summer's come and gone

And life just stumbles on

I wake up to the dawn wishing I was anywhere but here

I grab my freedom keys, I'm sick of this disease

I'm going where I please and I don't care if I disappear

(Give "This Town" by Rick Springfield a listen, and you'll see what I mean. And while you're at it, check out all of Rick's music and attend one or 20, 50, 100 of his concerts. I've lost count over the years, and I've also lost count of how many wonderful friends I've met because of Rick and his music.)

One of my greatest joys was taking my mom to a Rick Springfield concert in my hometown of Wausau, Wisconsin. Plus, I was blessed with the opportunity to take her backstage to meet Rick and chat a bit. I'm a writer, and I can't even find the words to express my heartfelt gratitude for that special moment in my life.

Family-loss

Relationships that do not strengthen you are better when lost. That brief escape to Maryland helped, but the divorce continued and brought the loss of some family relationships. I chose to stay loyal to those who were loyal to me. As for the family members who judged me and took a strong stance against my decision, though it was painful, I made the decision to move forward without them. They simply no longer had a place in my story.

Friend-loss

Like family, friends can sometimes feel caught in the middle of a divorce and feel like they have to choose sides. In my case, though, the loss of friendship was minimal. I'm incredibly grateful for my core groups of true friends—those who stood by me then and stand by me now and always. They're not just friends; they're my family, and their support means the world to me as I continue on my journey.

Some people come into our lives as blessings and lessons, and they may stay for just a chapter or the entire book. Someone who brings positivity, joy, support, or growth is a blessing. However, someone who teaches us valuable insights, often through difficult or challenging experiences, while helping us learn and grow as individuals, is usually a lesson. And most times, that lesson repeats itself until we've learned it. This I know for sure.

Home-loss

Through the separation and after the divorce, I worked hard to create a new home for myself and my girls, but they weren't exactly thrilled with our new place. It didn't feel like "home" to them. But it was my new reality, and it was the best I could offer them when we were together.

They always called our old house, now their dad's house "home," and I can understand why. It was the only home they'd ever known. I'd hear them say, I'm going to Mom's, not I'm going home too. Mom's place was not home to them—just a place to visit.

The divorce dragged on through guardians ad litem, attorneys, mediators, counselors, and countless uninvited opinions. Through it all, I stayed focused on healing and mending my relationship with my daughters. They were angry and blamed me for the divorce. They became my number one priority, as well as financially supporting myself, and owning my mistakes. However, waking up each day, knowing that precious bond I once shared with my daughters had been damaged was a pain I could hardly bear. I continued to pray to God for guidance every day.

"I am your rock and your salvation, a fortress that cannot be shaken" (Psalm 62:2).

A Mother's Love

So, who did I turn to first? Of course, my mom. Because sometimes, no matter how old you are, you just need your mom. And I love my mom; she is my greatest source of strength, my biggest cheerleader, and my best friend.

To know unconditional love is to know a mother's love, or this is the way it should be. I was still a good mom even though I wanted a divorce. I loved my daughters with all my heart. Children often believe their moms are invincible, but the truth is, moms are flawed, imperfect human beings. We're just trying to do our best, even when we fall short.

I had lost myself along the way, and I didn't know who I was anymore. All I wanted was to find my way back to peace, happiness, strength, self-sufficiency, and joy—and a positive role model who my daughters could look up to. I needed to earn their love again.

There's a scene in the 2015 film *Ricki and the Flash* that has always inspired me. Rick Springfield's character, Greg, says these words to Ricki, who is played by Meryl Streep, a mom who is struggling with the relationship with her grown children: "It doesn't matter if your kids love you. It's not their job to love you; it's your job to love them. That's why you were put here. That's why you're their mom."

These words couldn't ring more true. My one daughter who was like "my lost sheep" was angry at me, and at times didn't want to spend time together. I had to prove to her and my other daughter that I loved them no matter what. That was my job as their mom—to show them unconditional love.

"Which one of you, having a hundred sheep and losing one of them, does not leave the ninety-nine in the wilderness and go after the one that is lost until he finds it" (Luke 15:1-7).

My therapist helped me sort through my feelings of fear and loss, and she guided me through the understanding that my girls would find their way back to me. I needed to give them time, and I needed to continue to show my love through my words, and more importantly, my actions. They needed to see me as the same mom who stayed home with them through the years after they were born, walked them to and from the school bus, held them when they cried, comforted them when they couldn't sleep, went on school field trips, listened when they needed to talk or had exciting news to share, advocated for them when necessary, and loved them unconditionally.

After a lot of self-reflection work, focus, and determination, my girls and I slowly began to spend more time together and rebuild our relationship. They started trusting me again with their innermost thoughts as our mother/daughter bond grew even stronger.

God Keeps His Promises

My daughters eventually began attending my church with me, and before long, it felt like their church too. During the early days of the divorce, I had met with one of the pastors, and together we prayed for guidance, especially with "my lost sheep." We asked God to help us find our way back to each other, just as we once were. Over time, I continued to attend this church, growing closer to my faith, trusting that God answers our prayers—always in His time and according to His will.

"I hear you when you ask for anything according to My will" (1 John 5:14).

I love seeing how God works in wonderful and beautiful ways. Though it took years, my one daughter eventually became close friends with the pastor's daughter—the same pastor who had prayed for us years before. Through that friendship, she has met more great people in the church. It was a truly special moment when my girls and I gathered with the pastor to pray together, and I had the privilege of sharing that story with him. It felt like a full-circle moment of healing and grace.

"Trust me with all your heart, and I will guide you" (Proverbs 3:5-6).

The girls and I enjoyed and still enjoy our shopping days, sushi nights, movies, and of course, concerts. My daughters love concerts (just like me), so we'd crank up their favorite songs and jam out in the car during road trips up north and before heading to shows together. We made countless new memories while enjoying life's moments.

One of those special times was taking a trip to New York City. My mom joined the girls and me, and we experienced amazing sites and special moments together. We took in one of my daughter's favorite Broadway musicals, "Wicked." I immediately felt a strong bond of sisterhood during the show. I could see how my daughter's face lit up in the presence of this play, and I just wanted to experience that moment with her. We all loved it! She introduced me to the song "Defying Gravity," by Stephen Schwartz. The music and lyrics definitely resonated with me then and still resonate with me now:

"Defying Gravity"

Something has changed within me

Something is not the same

I'm through with playing by the rules of someone else's game

Too late for second-guessing

Too late to go back to sleep

It's time to trust my instincts, close my eyes and leap

It's time to try defying gravity

I think I'll try defying gravity

And you can't pull me down

✷✷✷

And if I'm flying solo

At least I'm flying free

To those who ground me

Take a message back from me

Tell them how I'm defying gravity

This song is about reclaiming your power, and I don't think it's a coincidence that as I struggled to write this chapter, the words began to flow more easily after I saw the recently released movie *Wicked*. Revisiting that song provided the inspiration I needed. The movie has had a profound impact on me, especially as I watch it through the eyes of my daughter.

"Home"

Two years ago, my daughters moved out of their dad's house and into their first apartment together. They were on their own for a year; however, working and attending school at the same time while trying to pay for life's expenses proved to be challenging. They asked if they could move in with me, but my two-bedroom apartment wasn't big enough for the three of us. That meant I had to move. I didn't give it a second thought, and we started looking for our new place to live.

Last summer, we moved into our new "home" together—just my daughters and me, plus one energetic dog and one sassy cat. As my girls have matured, I see them gaining a deeper understanding of life's complexities. They are my daughters, yes, but they are also young independent women carving out their place in this world—just as we all are. They both need to find their own pathways forward and move in the direction of their hopes and dreams. As their mom, I will do everything in my power to guide and encourage them along their journey. I wish them all of my love, joy, strength, hope, courage, and the belief in themselves to create the lives they envision. I love that we live together now—for however long that will be while they finish school. I've always tried to give them roots, but I'll need to let them go so they can find their wings and defy gravity on their own. I have found a refreshed peace and joy that was once lost, and I don't take any moments with them for granted. As they were, are, and will always be the brightest lights in all the chapters of my life.

The other day I could hear my daughter on the phone say, "I'm at Mom's, no … I mean I'm 'home.'"

Those were the sweetest words I've ever heard!

✷✷✷

For awhile, I struggled to write this chapter, and it ended up delaying the book release date. Usually, when I write, God gives me the words I need, but this time, it felt like He was silent—or perhaps I wasn't hearing Him clearly. One Sunday morning, as I sat in church listening to the sermon, the pastor said something that profoundly enlightened me: "The promise of God is always backed by the power of God." In that moment, I realized that God equips us with the power we need through His promises. I believe I needed to see more of His plan unfold before I could write the story He wanted me to share. I needed to see that by His power and through His promises, I would embrace my own power.

Proverbs 19:8 says, "Whoever gets sense loves his own soul; he who keeps understanding will discover good." When you embrace God's power and learn to love yourself, you reclaim your own strength. Self-love empowers you, ensuring that you don't give away your power to circumstances or others. Even though it felt like my life was falling apart at the time of my divorce and loss in my life, I came to understand that everything was actually falling into place.

By loving myself, I was able to take back my power. And, I know that nothing and no one will ever be able to take my power away from me again.

"And we know that in all things God works for the good of those who love him, who have been called according to his purpose" (Romans 8:28).

About Marla

Marla McKenna is a #1 bestselling author, speaker, editor, and graphic designer. While in school, Marla was involved in music, sports, and theater. She graduated from the University of Wisconsin receiving her Bachelor of Arts in Journalism. Marla's work is featured in these children's book: *Mom's Big Catch, Sadie's Big Steal, I'm a Secret Superhero*, and *A Soccer Summer Dream*. You can also find her books written for adults: *Manifesting Your Dreams—Inspiring Words of Encouragement, Strength, and Perseverance*, which reached #1 on Amazon and was featured in *O, The Oprah Magazine*. She is also the coauthor of *Our Last Day in Heaven, A Story of Tragedy, Loss, and Hope with Angels in the Midst*.

Her passion for writing and sharing a positive message with children comes to life in all of her children's books. Young readers were first introduced to Marla's heartwarming story *Mom's Big Catch*, which has given Marla the opportunity to partner with baseball teams and create customized team versions. She loves visiting schools and teaching the importance of patience, positivity, perseverance, and giving back while encouraging students to believe in themselves and never give up.

Marla's extensive work has appeared in magazines, brochures, newsletters, billboards, social media, radio spots, and websites. Her design work can also be seen in the Forbes bestselling book, *Culture Code Champions* and *A Powerful Promise*. She assisted in the ghostwriting efforts for a financial investment book and is the editor of several books written by Olympic athletes: Tianna Madison, Elizabeth Beisel, McKenzie Coan, and Katie Hoff. Marla also loves working with other writers and helping them get their work published.

Partial proceeds from all of Marla's books benefit the Linda Blair WorldHeart Foundation, with special thanks to Rick Springfield for matching her donations.

Marla is grateful for the loving relationships she has with her family and friends, while focusing on the positive influences in her life and giving back. Inspired by her daughters, Julia and Ashley, Marla continues to write and edit books while living in Wisconsin with her family. Watch for her new children's book, *Please Don't Make Me Choose*, a book about how children are affected by divorce.

For more information on Marla, please visit marlamckenna.com

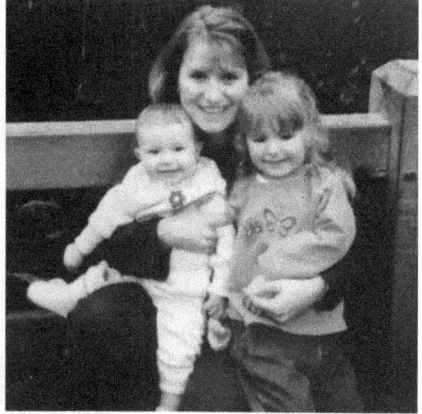

Marla and her daughters, Ashley and Julia, in California.

Marla, Ashley, and Julia.

Marla, Ashley, Julia, and Marla's mom, Marlene.

Marlene, Marla, Ashley, and Julia in California on Marla's birthday.

Marla's soul sister and friend, Penny, in Maryland at Rick Springfield's concert.

A fun selfie of Marla and Rick.

Marla, Rick, and Marlene backstage at his concert in Wausau, WI.

Marlene, Ashley, Julia, and Marla in New York City seeing "Wicked" on Broadway.

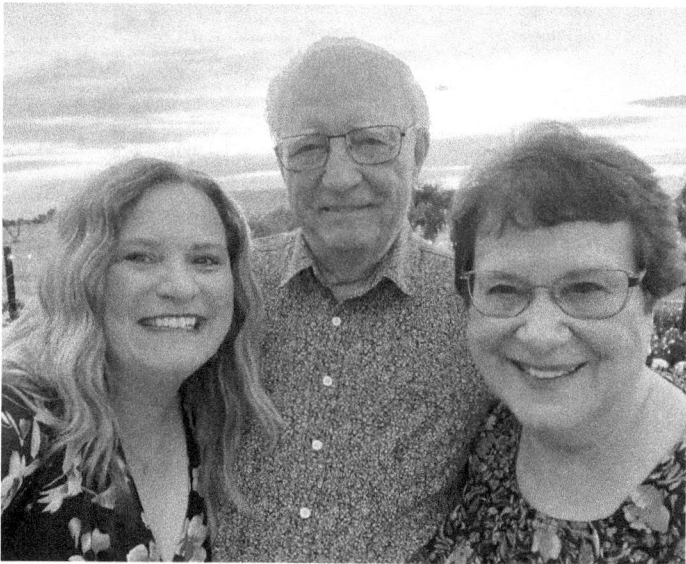

Marla and her parents, Jerry and Marlene, in California.

A gift to Marla from Julia and Ashley,
Marla's favorite flower is a sunflower.

Julia, Marla, and Ashley together at
the sunflower field. God is good.

"What lies behind us and what lies before us are tiny matters
compared to what lies within us."

- Ralph Waldo Emerson

A few words to remember (from this beautiful song) when you've lost your power and life isn't feeling very good. Don't lose hope. We are all here for you. You got this. Take back your power.

"Thy Will"

I know You're good
But this don't feel good right now
And I know you think
Of things I could never think about
It's hard to count it all joy
Distracted by the noise
Just tryna make sense
Of all Your promises
Sometimes I gotta stop
Remember that You're God
And I am not

So, Thy will be done
Thy will be done
Thy will be done like a child on my knees, all that comes to me is
Thy will be done
Thy will be done
Thy will
I know You see me
I know You hear me, Lord
Your plans are for me
Goodness You have in store
I know You hear me
I know You see me, Lord
Your plans are for me
Goodness You have in store
Thy will be done

Songwriters: Bernie Herms / Emily Lynn Weisband / Hillary Scott
Thy Will lyrics © Universal Music Publishing Group, Warner Chappell Music, Inc.